Learning Spring Boot 3.0

Third Edition

Simplify the development of production-grade applications
using Java and Spring

Greg L. Turnquist

BIRMINGHAM—MUMBAI

Learning Spring Boot 3.0

Third Edition

Copyright © 2022 Packt Publishing

Group Product Manager: Pavan Ramchandani

Publishing Product Manager: Bhavya Rao

Senior Editor: Mark D'Souza

Technical Editor: Simran Ali

Copy Editor: Safis Editing

Project Coordinator: Manthan Patel

Proofreader: Safis Editing

Indexer: Rekha Nair

Production Designer: Nilesh Mohite

Marketing Coordinator: Anamika Singh

First published: November 2014

Second Edition: November 2017

Third Edition: December 2022

Production reference: 2130123

Published by Packt Publishing Ltd.

Livery Place

35 Livery Street

Birmingham

B3 2PB, UK.

ISBN 978-1-80323-330-7

www.packt.com

To my fans on YouTube, who have given me the opportunity to share my love for Spring Boot. To my children, who have put up with me spending hours filming in my studio space. And to my wife, who has supported me in word and action as I've attempted to build a community.

– Greg L. Turnquist

Forewords

Spring Boot has been such a success that it's probably not wrong to describe it as "mainstream" in 2022. Practically every Java developer will know something about it, and many, maybe even the majority, will have used it, even if in anger. But in software engineering, there's always something new to learn, and there are always new problems to solve – that's what makes it so rewarding in the end. There's always something new to invent, too, and having the skill and opportunity to create code is extremely rewarding intellectually and in other ways.

One of the goals of Spring Boot is shared by the author of this book, and that is to get your ideas down into code as quickly and efficiently as possible, so you can get it to the most special place: Production. I wish you a short and pleasant journey, or maybe a long series of short and pleasant journeys.

In this book, Greg has used his insider advantage to add Spring Boot knowledge to some old, well-seasoned favourite problems that you all will have experienced as Java developers. What better way to learn than to look at Spring Boot through the lens of tasks that we all have to solve nearly every day: creating HTTP endpoints, securing them, connecting to databases, writing tests? This book adds some new angles to these old chestnuts by applying some modern ideas and tools, so read it and you will learn about things such as hypermedia and OAuth, all from the most practical and pragmatic of standpoints. It also starts right at the very beginning, and assumes no prior knowledge of Spring, or even Java. Anyone with some basic technical or programming skills will be able to get to grips with how and why to use Spring Boot.

There is more to Spring Boot than just main methods, embedded containers, autoconfiguration, and management endpoints. The pure joy of getting started with a fully featured Spring application in a few lines of code cannot be understated, for instance. I invite you to dip into this book, break out an editor or an IDE, and crank up some applications for yourself.

Greg has been an important member of the Spring Boot team, despite having a day job doing other things in the Spring Engineering effort, and we can be grateful for that, as well as the effort he has lavished on this excellent book. He has always been an educator and an informer, as well as an engineer, and this shows very clearly in this book. When I read it, I can hear Greg's voice and personality very clearly, and it is always calm but enthusiastic, with a touch of humour. Read it yourself and enjoy – coding with Spring is still fun after all these years!

Dave Syer

Sr. Staff Engineer and Co-Creator of Spring Boot

London, 2022

I've known Greg Turnquist for a number of years. After I joined Pivotal (and before I had officially started) I first met him in person at SpringOne2GX, the annual conference of all things Spring and more (anyone remember the heady days of Groovy and Grails?). We had some great, thought-provoking conversations there, and they've continued through the years since.

One of the first Spring Boot books I read was Greg's first edition of *Learning Spring Boot*. I wish I had read it sooner! I found myself recommending it to numerous people, along with other books by trusted colleagues, as a valuable introduction and reference to various Boot-related topics.

As an author myself, I know all too well the exhilarating and excruciating task Greg faced when writing and updating this book. Every author has to balance all the things they want to share, all the topics they feel most important, with the constant constraints of time and volume. Greg deftly threads this needle, providing a good foundation and then quickly shifting to topics important to developers, using Spring Boot to field real applications. Data? Security? Configuration? Integration with JavaScript? It's in there.

I loved working with Greg on the Spring team, and I continue to enjoy every conversation we have. There will always be an honored place on my (virtual) mantle for Greg's books, and I hope that you make room on yours for them as well. Read this book and get to know Greg! Your Spring Boot apps will benefit from both.

Best to you in your Spring Boot journey,

Mark Heckler

Principle Cloud Advocate, Microsoft

@mkheck

Look, I know you're probably reading this foreword hoping for some compelling testimonial about this book, along with some witticism and a fun anecdote about life. Why wouldn't you? It's a preface to a book. But I can't in good conscience write that foreword for you; it smacks of absurdity. But, of course, this book is fantastic. So, I don't want to linger on the obvious.

Let's talk about Greg, the author of this chef-d'oeuvre. Greg's been on the Spring team longer than I have been. He's forgotten more than most people will ever know about the depths of Spring. He invests time in the big *and* the small. You can trust him to be your sherpa and guide you from beginner to Spring Boot-iful. I do.

Greg's a friend. He and I get along because, in some crucial ways, we're very much alike. I like odd little projects that, while not necessarily mainstream, sometimes solve acutely painful problems. I once gave a talk to three people. My presentation was so specific that, out of thousands of attendees at the show, only three could be bothered to attend. I'm willing to advocate for the faintest glimmer of a solution if I believe in it. Greg is too. He invests in the big and small.

We both love the JVM *and* Python. That shared affection brings us around to Spring Python. Long ago, Greg brought some of the brilliance of the Spring Framework to the Python ecosystem with his project Spring Python. Python's ecosystem brims with alternatives for every use case. In this sea of choice, Spring Python stood out. It delivered on the lofty goals of the Spring Framework while remaining "Pythonic", a quality that signals a library will feel idiomatic to a familiar Python programmer. It showed a deep commitment to, and familiarity with, two vastly divergent ecosystems. I love Greg *because* of Spring Python. It shows he's willing to sit down, roll up his sleeves, expand his horizons, and write code until a problem is solved– no matter how big or small. That willingness to dive deep into a topic makes him a gifted writer and teacher, which is evident in his books, courses, blogs, and articles. His gift makes these printed pages something more than yet another book on software; they're a tome worth your time.

This book covers the just-released Spring Boot 3.0, arguably the most critical release of Spring Boot (or any other Spring ecosystem project) since Spring Boot itself appeared publicly for the very first time in 2013. I know all of us on the Spring team, Greg included, worked harder and longer than ever to get this release out the door. Yet, against all that work, Greg managed to get this book in your hands in record time. He did that so we, dear readers, could get to production in record time. He did that so we would *not* have to invest in the big and the small.

Josh Long

Spring Developer Advocate, VMware, (and well-known Greg Turnquist fan)

@starbuxman

Contributors

About the author

Greg L. Turnquist is the lead developer for Spring Data JPA and Spring Web Services. He has contributed to Spring HATEOAS, Spring Data REST, Spring Security, the Spring Framework, and many other parts of the Spring portfolio. He has maintained the Spring Data team's CI system for years with his script-fu. He has written multiple tomes on Spring Boot, including Packt's best-selling title *Learning Spring Boot 2.0 Second Edition* as well as the very first Spring Boot book to ever hit the market. He even launched his own YouTube channel, *Spring Boot Learning* (`http://bit.ly/3uSPLCz`), the channel where you learn about Spring Boot and have fun doing it. Before joining the Spring team, Greg worked as a senior software engineer at Harris Corp. on multiple projects including its ambitious $1.5 billion telecom contract with the FAA to build a nationwide, always-on network. As a test-bitten script junky, Greg is no stranger to midnight failures. He has a master's degree in computer engineering and lives in the United States with his wife and their gaggle of minions.

I want to thank the Spring team, who have encouraged me at every turn, Dan Vega, for giving me the inspiration to make YouTube content, and the team at Packt, who have worked tirelessly to help me publish this technical work.

About the reviewer

Harsh Mishra is a software engineer who enjoys learning new technologies for his own knowledge and experience, focusing on designing and developing enterprise solutions. He is a clean code and Agile fan. He has been developing code for financial businesses since 2014 and has been using Java as his primary programming language. He also has product experience in Spring, Microsoft, GCP, DevOps, and other enterprise technologies.

Table of Contents

Part 1: The Basics of Spring Boot

1

Core Features of Spring Boot   3

Part 2: Creating an Application with Spring Boot

2

Creating a Web Application with Spring Boot   23

Part 3: Releasing an Application with Spring Boot

6

Configuring an Application with Spring Boot   155

7

Releasing an Application with Spring Boot   171

8

Going Native with Spring Boot   189

Part 4: Scaling an Application with Spring Boot

9

Writing Reactive Web Controllers   203

10

Working with Data Reactively   227

Index   241

Other Books You May Enjoy   248

Preface

This book is designed for both novices and experienced Spring developers. It will teach you how to build Java applications without wasting time on infrastructure and other tedious details. Instead, it will help you focus on building web apps on top of real databases, locked down with modern security practices.

On top of that, you'll discover multiple ways to carry an app to production. If that's not enough, it even includes secret ways (okay, not really secret) at the end to squeeze more out of your existing servers (or cloud) by picking up and running with reactive programming.

Who this book is for

This book is designed for both novices and experienced Spring developers who want to get their hands on Spring Boot 3.0. You should have a rudimentary understanding of Java, preferably Java 8 or higher. Being familiar with lambda functions, method references, record types, and the new-and-improved collections API found in Java 17 is a bonus, but not required.

Having used previous versions of Spring Boot (1.x, 2.x) is not required but would be handy.

What this book covers

Chapter 1, Core Features of Spring Boot, is where you discover the charm of Spring Boot with its fundamental features to work with you as you build your app.

Chapter 2, Creating a Web Application with Spring Boot, teaches you how to craft the web layer for a Java app with ease, with both server-side and client-side options.

Chapter 3, Querying for Data with Spring Boot, shows you how to get the most out of your database with Spring Data.

Chapter 4, Securing an Application with Spring Boot, shows you how to use Spring Security's cutting-edge features to lock down your app from bad guys, inside and out.

Chapter 5, Testing with Spring Boot, teaches you how to build confidence in your systems through testing with mocks and embedded databases, and even using Testcontainers combined with real databases.

Chapter 6, Configuring an Application with Spring Boot, shows you ways to tune and adapt your app once it's built.

Chapter 7, Releasing an Application with Spring Boot, helps you discover multiple ways to take your app to production and put it in the hands of your users.

Chapter 8, Going Native with Spring Boot, shows you how to speed up your app by leaps and bounds using native images that start in subsecond time and don't hog all your resources.

Chapter 9, Writing Reactive Web Controllers, teaches you how easy it is to write reactive web controllers and how they can be your key to a more efficient app.

Chapter 10, Working with Data Reactively, helps you discover how to query for data reactively using R2DBC and see how efficient your app can be.

To get the most out of this book

Spring Boot 3.0 is built on Java 17. By using sdkman (`https://sdkman.io`), you can easily install the version of Java needed. In *Chapter 8, Going Native with Spring Boot*, we'll include instructions on how to use sdkman to install a specific flavor of Java 17 that supports building native images for GraalVM. While it's possible to write code using a barebones text editor, any modern IDE (see the following list) will greatly enhance the coding experience. Check out the one that works best for you.

Software/hardware covered in the book	Operating system requirements
sdkman (for Java 17) (`https://sdkman.io`)	Windows, macOS, or Linux
Any modern IDE will help with writing code: • IntelliJ IDEA (`https://springbootlearning.com/intellij-idea-try-it`) • VS Code (`https://springbootlearning.com/vscode`) • Spring Tool Suite (`https://springbootlearning.com/sts`)	Windows, macOS, or Linux

VS Code and Spring Tool Suite are free. IntelliJ IDEA has a Community Edition and an Ultimate Edition. The Community Edition is free, but some of the Spring-specific features require purchasing the Ultimate Edition. There is a 30-day trial to give it a spin.

If you are using the digital version of this book, we advise you to type the code yourself or access the code from the book's GitHub repository (a link is available in the next section). Doing so will help you avoid any potential errors related to the copying and pasting of code.

This book, however, isn't the end of your journey into building Spring Boot apps. Check out my YouTube channel, *Spring Boot Learning* (`http://bit.ly/3uSPLCz`), where I publish videos all the time on Spring Boot and software engineering. There are also additional resources at `https://springbootlearning.com` to help you write better apps!

Download the example code files

You can download the example code files for this book from GitHub at `https://github.com/ PacktPublishing/Learning-Spring-Boot-3.0`. If there's an update to the code, it will be updated in the GitHub repository.

We also have other code bundles from our rich catalog of books and videos available at `https://github.com/PacktPublishing/`. Check them out!

Download the color images

We also provide a PDF file that has color images of the screenshots and diagrams used in this book. You can download it here: `https://packt.link/FvE6S`.

Conventions used

There are a number of text conventions used throughout this book.

`Code in text`: Indicates code words in text, database table names, folder names, filenames, file extensions, pathnames, dummy URLs, user input, and Twitter handles. Here is an example: "This can be done by first adding an `application.properties` file to our `src/main/resources` folder."

A block of code is set as follows:

```
@Controller
public class HomeController {

  private final VideoService videoService;

  public HomeController(VideoService videoService) {
    this.videoService = videoService;
  }

  @GetMapping("/")
  public String index(Model model) {
    model.addAttribute("videos", videoService.getVideos());
    return "index";
  }
}
```

When we wish to draw your attention to a particular part of a code block, the relevant lines or items are set in bold:

```
@Bean
SecurityFilterChain configureSecurity(HttpSecurity http) {
  http.authorizeHttpRequests()
    .requestMatchers("/login").permitAll()
    .requestMatchers("/", "/search").authenticated()
    .anyRequest().denyAll()
    .and()
    .formLogin()
    .and()
    .httpBasic();
  return http.build();
}
```

Any command-line input or output is written as follows:

```
$ cd ch7
$ ./mvnw clean spring-boot:build-image
```

Bold: Indicates a new term, an important word, or words that you see onscreen. For instance, words in menus or dialog boxes appear in **bold**. Here is an example: "To do that, go to the **Dependencies** section."

> **Tips or important notes**
> Appear like this.

Get in touch

Feedback from our readers is always welcome.

General feedback: If you have questions about any aspect of this book, email us at customercare@ packtpub.com and mention the book title in the subject of your message.

Errata: Although we have taken every care to ensure the accuracy of our content, mistakes do happen. If you have found a mistake in this book, we would be grateful if you would report this to us. Please visit www.packtpub.com/support/errata and fill in the form.

Piracy: If you come across any illegal copies of our works in any form on the internet, we would be grateful if you would provide us with the location address or website name. Please contact us at copyright@packt.com with a link to the material.

If you are interested in becoming an author: If there is a topic that you have expertise in and you are interested in either writing or contributing to a book, please visit authors.packtpub.com.

Share Your Thoughts

Once you've read *Learning Spring Boot 3.0*, we'd love to hear your thoughts! Scan the QR code below to go straight to the Amazon review page for this book and share your feedback.

https://packt.link/r/1803233303

Your review is important to us and the tech community and will help us make sure we're delivering excellent quality content.

Download a free PDF copy of this book

Thanks for purchasing this book!

Do you like to read on the go but are unable to carry your print books everywhere?

Is your eBook purchase not compatible with the device of your choice?

Don't worry, now with every Packt book you get a DRM-free PDF version of that book at no cost.

Read anywhere, any place, on any device. Search, copy, and paste code from your favorite technical books directly into your application.

The perks don't stop there, you can get exclusive access to discounts, newsletters, and great free content in your inbox daily

Follow these simple steps to get the benefits:

1. Scan the QR code or visit the link below

https://packt.link/free-ebook/9781803233307

2. Submit your proof of purchase
3. That's it! We'll send your free PDF and other benefits to your email directly

Part 1: The Basics of Spring Boot

Spring Boot has several key ingredients that underpin all of its features. You will learn how autoconfiguration, Spring Boot starters, configuration properties, and managed dependencies make it possible to build your most powerful application yet.

This part includes the following chapter:

- *Chapter 1, Core Features of Spring Boot*

Core Features of Spring Boot

Rod Johnson, CEO of the company behind the foundation of the **Spring Framework** and dubbed the *father of Spring*, opened the 2008 Spring Experience conference with a stated mission: reducing Java complexity. The YouTube video titled *Story time with Keith Donald Co-Founder SpringSource & Founder SteadyTown 2-27-2014* (`https://springbootlearning.com/origin-of-spring`), uploaded by *TrepHub*, is a 90-minute journey back into the early days of Spring guided by Keith Donald, one of Spring's co-founders. Here too, you'll find the same mission reinforced.

Java in the mid-2000s was challenging to use, difficult to test, and frankly short on enthusiasm.

But along came a toolkit: the Spring Framework. This toolkit focused on easing developers' lives. And the excitement was off the charts. The buzz when I attended that 2008 conference was incredible.

Fast forward to 2013 at the *SpringOne 2GX conference*, the Spring team unveiled **Spring Boot**: a new approach to writing Spring apps. This approach resulted in standing-room attendance. I was in the room when co-leads Phil Webb and Dave Syer gave their first talk. In a room designed like a stadium lecture hall, the seats were packed. The opening keynote revealed a revolutionary way to build more extensive and powerful apps… with less.

This ability to do more with less using Spring Boot is what we'll discover together as we journey into the world of the third generation of Spring Boot.

In this chapter, we'll learn about the core features of Spring Boot, which show fundamentally how it does less with more. This is to get a taste of how Spring Boot operates, allowing us to leverage it in later chapters as we build applications. The key aspects that make Spring Boot powerful while retaining its flexibility to serve user needs will be described in this chapter.

In this chapter, we'll cover the following topics:

- Autoconfiguring Spring beans
- Adding components of the Spring portfolio using Spring Boot starters
- Customizing the setup with configuration properties
- Managing application dependencies

Technical requirements

For this book, you'll only need a handful of tools to follow along:

- **Java 17 Development Kit (JDK 17)**
- A modern **integrated development environment (IDE)**
- A GitHub account
- Additional support

Installing Java 17

Spring Boot 3.0 is built on top of Java 17. For ease of installation and using Java, it's easiest to use **sdkman** as your tool to handle installing and switching between different JDKs, as shown here:

1. Visit `https://sdkman.io/`.
2. Following the site's instructions, execute `curl -s "https://get.sdkman.io" | bash` from any terminal or shell on your machine.
3. Follow any subsequent instructions provided.
4. From there, install Java 17 on your machine by typing `sdk install java 17.0.2-tem`. When prompted, pick it as your default JDK of choice in any terminal.

This will download and install the **Eclipse Temurin** flavor of Java 17 (formerly known as **AdoptOpenJDK**). Eclipse Temurin is a free, open source version of OpenJDK, compliant with all standard Java TCKs. In general, it's a variant of Java recognized by all parties as acceptable for Java development. Additionally, it comes with no requirements to pay for licensing.

> **Tip**
>
> If you need a commercially supported version of Java, then you will have to do more research. Many shops that provide commercial support in the Java space will have various options. Use what works best for you. But if commercial support is not needed, then Eclipse Temurin will work fine. It's used by many projects managed by the Spring team itself.

Installing a modern IDE

Most developers today use one of the many free IDEs to do their development work. Consider these options:

- IntelliJ IDEA – Community Edition (`https://www.jetbrains.com/idea/`)

- Spring Tools 4 (`https://spring.io/tools`):

  - Spring Tools 4 for Eclipse

  - Spring Tools 4 for VS Code

IntelliJ IDEA is a powerful IDE. The Community Edition, which is free, has many bits that will serve you well. The Ultimate Edition, which costs 499 USD, is a complete package. If you grab this (or convince your company to buy a license!), it's a valuable investment.

That being said, Spring Tools 4, whether you pick the Eclipse flavor or the VS Code one, is a powerful combo as well.

If you're not sure, go ahead and test out each one, perhaps for a month, and see which one provides you with the best features. They each have top-notch support for Spring Boot.

At the end of the day, some people do prefer a plain old text editor. If that's you, fine. At least evaluate these IDEs to understand the tradeoffs.

Creating a GitHub account

I always tell anyone entering the world of 21st century software development to open a GitHub account if they haven't already. It will ease access to so many tools and systems out there.

Visit `https://github.com/join` if you're just getting started.

This book's code is hosted on GitHub at `https://github.com/PacktPublishing/Learning-Spring-Boot-3.0`.

You can work your way through the code presented in this book, but if you need to go to the source, visit the aforementioned link and grab a copy for yourself!

Finding additional support

Finally, there are some additional resources to visit for more help:

- I host a YouTube channel focused on helping people get started with Spring Boot at `https://youtube.com/@SpringBootLearning`. All the videos and live streams there are completely free.

- There is additional content provided to my exclusive members at `https://springbootlearning.com/member`. My members also get one-on-one access to me with questions and concerns.

- If you're a paying subscriber on Medium, I also write technical articles based on Spring Boot, along with overall software development topics, at `https://springbootlearning.medium.com`. Follow me over there.

- I also share any technical articles posted with my newsletter at `https://springbootlearning.com/join` for free. You also get an e-book for free if you sign up.

If you've downloaded Java 17 and installed an IDE, then you're all set, so let's get to it!

Autoconfiguring Spring beans

Spring Boot comes with many features. But the most well-known one, by far, is autoconfiguration.

In essence, when a Spring Boot application starts up, it examines many parts of our application, including `classpath`. Based on what the application sees, it automatically adds additional **Spring beans** to the **application context**.

Understanding application context

If you're new to Spring, then it's important to understand what we're talking about when you hear application context.

Whenever a Spring Framework application starts up, whether or not Spring Boot is involved, it creates a container of sorts. Various Java beans that are registered with Spring Framework's application context are known as Spring beans.

> **Tip**
>
> What's a Java bean? Java beans are objects that follow a specific pattern: all the fields are private; they provide access to their fields through getters and setters, they have a no-argument constructor, and they implement the `Serializable` interface.
>
> For example, an object of the `Video` type with `name` and `location` fields would set those two fields to `private` and offer `getName()`, `getLocation()`, `setName()`, and `setLocation()` as the ways to mutate the state of this bean. On top of that, it would have a no-argument `Video()` constructor call. It's mostly a convention. Many tools provide property support by leveraging the getters and setters. The requirement to implement the `Serializable` interface, though, is not as tightly enforced.

Spring Framework has a deep-seated concept known as **dependency injection (DI)**, where a Spring bean can express its need for a bean of some other type. For example, a `BookRepository` bean may require a `DataSource` bean:

```
@Bean
public BookRepository bookRepository(DataSource dataSource) {
    return new BookRepository(dataSource);
}
```

This preceding Java configuration, when seen by the Spring Framework, will cause the following flow of actions:

1. `bookRepository` needs a `DataSource`.
2. Ask the application context for a `DataSource`.
3. The application context either has it or will go create one and return it.
4. `bookRepository` executes its code while referencing the app context's `DataSource`.
5. `BookRepository` is registered in the application context under the name `bookRepository`.

The application context will ensure all Spring beans needed by the application are created and properly injected into each other. This is known as *wiring*.

Why all this instead of a handful of new operations in various class definitions? Simple. For the standard situation of powering up our app, all the beans are wired together as expected.

For a test case, it's possible to override certain beans and switch to stubbed or mocked beans.

For cloud environments, it's easy to find all `DataSource` and replace them with beans that link to bound data services.

By removing the new operation from our example `BookRepository`, and delegating that responsibility to the application context, we open the door to flexible options that make the whole life cycle of application development and maintenance much easier.

We'll explore how Spring Boot heavily leverages the Spring Framework's ability to inject beans based on various circumstances throughout this book. It is important to realize that Spring Boot doesn't replace the Spring Framework but rather highly leverages it.

Now that you know what an application context is, it is time to dive into the many ways Spring Boot makes use of it through autoconfiguration.

Exploring autoconfiguration policies in Spring Boot

Spring Boot comes with a fistful of autoconfiguration policies. These are classes that contain `@Bean` definitions that are only registered based on certain conditional circumstances. Perhaps an example is in order?

If Spring Boot detects the class definition of `DataSource` somewhere on the `classpath`, a class found inside any **Java Database Connectivity (JDBC)** driver, it will activate its `DataSourceAutoConfiguration`. This policy will fashion some version of a `DataSource` bean. This is driven by the `@ConditionalOnClass({ DataSource.class })` annotation found on that policy.

Inside `DataSourceAutoConfiguration` are inner classes, each driven by various factors. For example, some classes will discern whether or not we have used an embedded database such as **H2** compared to a pooled JDBC asset such as `HikariCP`.

And just like that, the need for us to configure an H2 `DataSource` is removed. A small piece of infrastructure that is often the same across a multitude of applications is taken off our plate and instead managed by Spring Boot. And we can move more quickly toward writing business code that uses it.

Spring Boot autoconfiguration also has smart ordering built in, ensuring beans are added properly. Don't worry! Using Spring Boot doesn't depend on us having to know this level of detail.

Most of the time, we don't have to know what Spring Boot is up to. It's designed to do the right thing when various things are added to the build configuration.

The point is that many features, such as servlet handlers, view resolvers, data repositories, security filters, and more are activated, simply based on what dependencies we add to the build file.

And do you know what's even better than automagically adding Spring beans? Backing off.

Some beans are created based on the `classpath` settings. But if a certain bean definition is detected inside our code, the autoconfiguration won't kick in.

Continuing with the example from earlier, if we put something such as H2 in our `classpath` but define a `DataSource` bean and register it in the application context, Spring Boot will accept our `DataSource` bean over theirs.

No special hooks. No need to tell Spring Boot about it. Just create your own bean as you see fit, and Spring Boot will pick it up and run with it!

This may sound low-level, but Spring Boot's autoconfiguration feature is transformational. If we focus on adding all the dependencies our project needs, Spring Boot will, as stated earlier, do what's right.

Some of the autoconfiguration policies baked into Spring Boot extend across these areas:

- **Spring AMQP**: Communicate asynchronously using an **Advanced Message Queueing Protocol (AMQP)** message broker
- **Spring AOP**: Apply advice to code using **Aspect-Oriented Programming**
- **Spring Batch**: Process large volumes of content using batched jobs
- **Spring Cache**: Ease the load on services by caching results
- **Data store connections** (Apache Cassandra, Elasticsearch, Hazelcast, InfluxDB, JPA, MongoDB, Neo4j, Solr)
- **Spring Data** (Apache Cassandra, Couchbase, Elasticsearch, JDBC, JPA, LDAP, MongoDB, Neo4j, R2DBC, Redis, REST): Simplify data access
- **Flyway**: Database schema management

- **Templating engines** (Freemarker, Groovy, Mustache, Thymeleaf)

- **Serialization/deserialization** (Gson and Jackson)

- **Spring HATEOAS**: Add **Hypermedia as the Engine of Application State** (HATEOAS) or hypermedia to web services

- **Spring Integration**: Support integration rules

- **Spring JDBC**: Simplify accessing databases through **JDBC**

- **Spring JMS**: Asynchronous through **Java Messaging Service** (JMS)

- **Spring JMX**: Manage services through **Java Management Extension** (JMX)

- **jOOQ**: Query databases using **Java Object Oriented Querying** (jOOQ)

- **Apache Kafka**: Asynchronous messaging

- **Spring LDAP**: Directory-based services over **Lightweight Directory Access Protocol** (jOOQ)

- **Liquibase**: Database schema management

- **Spring Mail**: Publish emails

- **Netty**: An asynchronous web container (non-servlet-based)

- **Quartz scheduling**: Timed tasks

- **Spring R2DBC**: Access relational databases through **Reactive Relational Database Connectivity (R2DBC)**

- **SendGrid**: Publish emails

- **Spring Session**: Web session management

- **Spring RSocket**: Support for the async wire protocol known as RSocket

- **Spring Validation**: Bean validation

- **Spring MVC**: Spring's workhorse for servlet-based web apps using the **Model-View-Controller** (**MVC**) paradigm

- **Spring WebFlux**: Spring's reactive solution for web apps

- **Spring Web Services**: **Simple Object Access Protocol** (SOAP)-based services

- **Spring WebSocket**: Support for the WebSocket messaging web protocol

This is a general list and is by no means exhaustive. It's meant to give us a glance at the breadth of Spring Boot.

And as cool as this set of policies and its various beans are, it's lacking a few things that would make it perfect. For example, can you imagine managing the versions of all those libraries? And what about hooking in our own settings and components? We'll cover these aspects in the next few sections.

Adding portfolio components using Spring Boot starters

Remember in the previous section how we talked about adding H2? Or **Spring MVC**? Maybe **Spring Security**?

I'm going out on a limb here, but I'm presuming you don't have any project dependency coordinates committed to memory. What does Spring Boot offer? A collection of virtual dependencies that will ease adding things to the build.

If you add `org.springframework.boot:spring-boot-starter-web` (as shown here) to the project, it will activate Spring MVC:

```
<dependency>
    <groupId>org.springframework.boot</groupId>
    <artifactId>spring-boot-starter-web</artifactId>
</dependency>
```

If you add `org.springframework.boot:spring-boot-starter-data-jpa` to the project (as shown here), it will activate Spring Data JPA:

```
<dependency>
    <groupId>org.springframework.boot</groupId>
    <artifactId>spring-boot-starter-data-jpa</artifactId>
</dependency>
```

There are 50 different Spring Boot starters, each with the perfect coordinates to various bits of the Spring portfolio and other related third-party libraries.

But the problem isn't just a shortcut to adding Spring MVC to `classpath`. There's little difference between `org.springframework.boot:spring-boot-starter-web` and `org.springframework:spring-webmvc`. That's something we probably could have figured out with our favorite internet search engine.

No, the issue is that if we want Spring MVC, it implies we probably want the whole **Spring Web** experience.

> **Note**
>
> **Spring MVC** versus **Spring Web**? Spring Framework has three artifacts involving web applications: Spring Web, Spring MVC, and Spring WebFlux. Spring MVC is servlet-specific bits. Spring WebFlux is for reactive web app development and is not tied to any servlet-based contracts. Spring Web contains common elements shared between Spring MVC and Spring WebFlux. This mostly includes the annotation-based programming model Spring MVC has had for years. This means that the day you want to start writing reactive web apps, you don't have to learn a whole new paradigm to build web controllers.

If we added `spring-boot-starter-web`, this is what we'd need:

- Spring MVC and the associated annotations found in Spring Web. These are the Spring Framework bits that support servlet-based web apps.

- Jackson Databind for serialization and deserialization (including JSR 310 support) to and from JSON.

- An embedded Apache Tomcat servlet container.

- Core Spring Boot starter.

- Spring Boot.

- Spring Boot Autoconfiguration.

- Spring Boot Logging.

- Jakarta annotations.

- Spring Framework Core.

- SnakeYAML to handle **YAML Ain't Markup Language** (**YAML**)-based property files.

> **Note**
>
> What is **Jakarta**? **Jakarta EE** is the new official specification, replacing **Java EE**. Oracle wouldn't relinquish its trademarked Java brand (nor grant a license) when it released its Java EE specs to the **Eclipse Foundation**. So, the Java community chose *Jakarta* as the new brand going forward. Jakarta EE 9+ is the official version that Spring Boot 3.0 supports. For more details, checkout my video *What is Jakarta EE?* at `https://springbootlearning.com/jakarta-ee`

This starter will bring in enough for you to build a real web application, not counting a templating engine. Now, we have autoconfiguration, which registers key beans in the application context. And we also have starters that simplify putting Spring portfolio components in `classpath`. But what's missing is our ability to plug in customized settings, which we'll tackle in the next section.

Customizing the setup with configuration properties

So, we've decided to pick up Spring Boot and we started adding some of its magical starters. As discussed earlier in this chapter, this will activate a handful of Spring beans.

Assuming we were building a web app and selected Spring MVC's `spring-boot-starter-web`, it would activate embedded **Apache Tomcat** as the servlet container of choice. And with that, Spring Boot is forced to make a lot of assumptions.

For example, what port should it listen on? What about the context path? **Secure Sockets Layer** (**SSL**)? Threads? There are a dozen other parameters to fire up a Tomcat servlet container.

And Spring Boot will pick them. So, where does that leave us? Are we stuck with them? No.

Spring Boot introduces **configuration properties** as a way to plug property settings into any Spring bean. Spring Boot may load certain properties with default values, but we have the opportunity to override them.

The simplest example is the first property mentioned earlier in this section – the server port.

Spring Boot launches with a default port in mind, but we can change it. This can be done by first adding an `application.properties` file to our `src/main/resources` folder. Inside that file, we must merely add the following:

```
server.port=9000
```

This Java property file, a file format supported since the early days of Java 1.0, contains a list of key-value pairs separated by an equals sign (`=`). The left-hand side contains the key (`server.port`) and the right-hand side contains the value (`9000`).

When a Spring Boot application launches, it will look for this file and scan in all its property entries, and then apply them. And with that, Spring Boot will switch from its default port of `8080` to port `9000`.

> **Note**
> The server port property is really handy when you need to run more than one Spring Boot-based web application on the same machine.

Spring Boot is not limited to the handful of properties that can be applied to embedded with Apache Tomcat. Spring Boot has alternative servlet container starters, including *Jetty* and *Undertow*. We'll learn how to pick and choose servlet containers in *Chapter 2, Creating a Web Application with Spring Boot*.

What's important is knowing that no matter which servlet container we use, the `servlet.port` property will be applied properly to switch the port the servlet will serve web requests on.

Perhaps you're wondering why? Having a common port property between servlet containers eases choosing servlet containers.

Yes, there are container-specific property settings if we needed that level of control. But generalized properties make it easy for us to select our preferred container and move to a port and context path of choice.

But we're getting ahead of ourselves. The point of Spring Boot property settings isn't about servlet containers. It's about creating opportunities to make our applications flexible at runtime. And the next section will show us how to create configuration properties.

Creating custom properties

At the beginning of this section, I mentioned that configuration properties can be applied to any Spring bean. This applies not just to Spring Boot's autoconfigured beans, but to our own Spring beans!

Look at the following code:

```
@Component
@ConfigurationProperties(prefix = "my.app")
public class MyCustomProperties {
    // if you need a default value, assign it here or the
        constructor
    private String header;
    private String footer;

    // getters and setters
}
```

The preceding code can be described as follows:

- @Component is Spring Framework's annotation to automatically create an instance of this class when the application starts and register it with the application context.
- @ConfigurationProperties is a Spring Boot annotation that labels this Spring bean as a source of configuration properties. It indicates that the prefix of such properties will be my.app.

The class itself must adhere to standard Java bean property rules (described earlier in this chapter). It will create various fields and include proper getters and setters – in this case, getHeader() and getFooter().

With this class added to our application, we can include our own custom properties, as follows:

```
application.properties:
my.app.header=Learning Spring Boot 3
my.app.footer=Find all the source code at https://github.com/
PacktPublishing/Learning-Spring-Boot-3.0
```

These two lines will be read by Spring Boot and injected into the MyCustomProperties Spring bean before it gets injected into the application context. We can then inject that bean into any relevant component in our app.

But a much more tangible concept would be including properties that should never be hardcoded into an application, as follows:

```
@Component
@ConfigurationProperties(prefix = "app.security")
public class ApplicationSecuritySettings {

  private String githubPersonalCode;

  public String getGithubPersonalCode() {
    return this.githubPersonalCode;
  }

  public void setGithubPersonalCode
    (String githubPersonalCode) {
      this.githubPersonalCode = githubPersonalCode;
  }
}
```

The preceding code is quite similar to the earlier code but with the following differences:

- The prefix for this class's properties is `app.security`
- The `githubPersonalCode` field is a string used to store an API passcode used to presumably interact with GitHub through its OAuth API

An application that needs to interact with GitHub's API will need a passcode to get in. We certainly do not want to bake that into the application. What if the passcode were to change? Should we rebuild and redeploy the whole application just for that?

No. It's best to delegate certain aspects of an application to an external source. How can we do that? The next section will show how!

Externalizing application configuration

Did I mention an external source in the previous section? Yes. That's because while you can put properties into an `application.properties` file that gets baked into the application, that isn't the only way to do things. There are more options when it comes to providing Spring Boot with application properties that aren't solely inside the deliverable.

Spring Boot not only looks for that `application.properties` tucked inside our JAR file upon startup. It will also look directly in the folder from where we run the application to find any `application.properties` files there and load them.

We can deliver our JAR file along with an `application.properties` file right beside it as an immediate way to override pre-baked properties (ours or Spring Boot's!).

But wait, there's more. Spring Boot also supports **profiles**.

What are profiles? We can create profile-specific property overrides. A good example would be one configuration for the development environment, but a different one for our test bed, or production.

In essence, we can create variations of `application.properties`, as shown here:

- `application-dev.properties` is a set of properties applied when the `dev` profile is activated
- `application-test.properties` is applied when the `test` profile is applied
- `application.properties` is always applied, so it could be deemed the production environment

Perhaps an example is in order?

Imagine having our database connection details captured in a property named `my.app.databaseUrl`, as shown here:

```
application.properties:
my.app.databaseUrl=https://user:pass@production-server.
com:1234/prod/
```

The test bed of our system surely won't be linked to the same production server. So, instead, we must provide an `application-test.properties` with the following override:

```
application-test.properties:
my.app.databaseUrl=http://user:pass@test-server.com:1234/test/
```

To activate this override, simply include `-Dspring.profiles.active=test` as an extra argument to the Java command to run our app.

It's left as an exercise for you to think up overrides for a development environment.

> **Note**
> Since production is the end state of an application, it's usually best practice to let `application.properties` be the production version of property settings. Use other profiles for other environments or configurations.

Notice earlier how we said Spring Boot will scan either `application.properties` files embedded inside our JAR as well as outside the JAR? The same goes for profile-specific property files.

So far, we've mentioned internal and external properties, both default and profile-specific. In truth, there are many more ways to bind property settings into a Spring Boot application.

Several are included in this list, ordered from lowest priority to highest priority:

- Default properties provided by Spring Boot's `SpringApplication.setDefaultProperties()` method.

- `@PropertySource`-annotated `@Configuration` classes.

- Config data (such as `application.properties` files).

- A `RandomValuePropertySource` that has properties only in `random.*`.

- OS environment variables.

- Java system properties (`System.getProperties()`).

- JNDI attributes from `java:comp/env`.

- `ServletContext` init parameters.

- `ServletConfig` init parameters.

- Properties from `SPRING_APPLICATION_JSON` (inline JSON embedded in an environment variable or system property).

- Command-line arguments.

- The `properties` attribute on your tests. This is available with the `@SpringBootTest` annotation and also slice-based testing (which we'll cover later in *Chapter 5, Testing with Spring Boot*).

- `@TestPropertySource` annotations on your tests.

- DevTools global settings properties (the `$HOME/.config/spring-boot` directory when Spring Boot DevTools is active).

Config files are considered in the following order:

- Application properties packaged inside your JAR file.

- Profile-specific application properties inside your JAR file.

- Application profiles outside your JAR file.

- Profile-specific application properties outside your JAR file.

It's a bit of a tangent, but we can also ensure certain beans are *only* activated when certain profiles are activated.

And properties aren't confined to injecting data values. The following section will show you how to make property-based beans.

Configuring property-based beans

Properties aren't just for providing settings. They can also govern which beans are created and when.

The following code is a common pattern for defining beans:

```
@Bean
@ConditionalOnProperty(prefix="my.app", name="video")
YouTubeService youTubeService() {
    return new YouTubeService();
}
```

The preceding code can be explained as follows:

- @Bean is Spring's annotation, signaling that the following code should be invoked when creating an application context and the created instance is added as a Spring bean
- @ConditionalOnProperty is Spring Boot's annotation to conditionalize this action based on the existence of the property

If we set my.app.video=youtube, then a bean of the YouTubeService type will be created and injected into the application context. Actually, in this scenario, if we define my.app.video with any value, it will create this bean.

If the property does not exist, then the bean won't be created. This saves us from having to deal with profiles.

It's possible to fine-tune this even further, as shown here:

```
@Bean
@ConditionalOnProperty(prefix="my.app", name="video",
havingValue="youtube")
YouTubeService youTubeService() {
    return new YouTubeService();
}
@Bean
@ConditionalOnProperty(prefix="my.app", name="video",
havingValue="vimeo")
VimeoService vimeoService() {
    return new VimeoService();
}
```

This preceding code can be explained as follows:

- `@Bean`, like before, will define Spring beans to be created and added to the application context
- `@ConditionalOnProperty` will conditionalize these beans to only be created if the named property has the stated values

This time, if we set `my.app.video=youtube`, a `YouTubeService` will be created. But if we were to set `my.app.video=vimeo`, a `VimeoService` bean would be created instead.

All of this presents a rich way to define application properties. We can create all the configuration beans we need. We can apply different overrides based on various environments. And we can also conditionalize which variants of various services are created based on these properties.

We can also control which property settings are applicable in a given environment, be it a test bed, a developer's work environment, a production setting, or a backup facility. We can even apply additional settings based on being in different cloud providers!

And as a bonus, most modern IDEs (IntelliJ IDEA, Spring Tool Suite, Eclipse, and VS Code) offer autocompletion inside `application.properties` files! We will cover this in more detail throughout the rest of this book.

Now, the last thing we need to craft a powerful application is the means to maintain it. This will be covered in the next section.

Managing application dependencies

There's a subtle thing we may have glossed over. It is best expressed in this simple question:

What version of Spring Framework works best with which version of Spring Data JPA and Spring Security?

Indeed, that is quite tricky. In fact, over the years, thousands of hours have probably been spent simply managing version dependencies.

Imagine that a new version of Spring Data JPA is released. It has an update to the *Query by Example* option you've been waiting for – the one where they finally handle domain objects that use *Java's Optional type* in the getters. It's been bugging you because anytime you had an `Optional.EMPTY`, it just blew up.

So, you're eager to upgrade.

But you don't know if you can. The last upgrade cost you a week of effort. It included digging through release reports for Spring Framework as well as Spring Data JPA.

Your system also uses **Spring Integration** and Spring MVC. If you bump up the version, will those other dependencies run into any issues?

With the miracle of autoconfiguration, all those slick starters and easy-to-use configuration properties can come off as a bit weak if you're left dealing with this conundrum.

That's why Spring Boot also comes loaded with an extensive list of 195 approved versions. If you pick a version of Spring Boot, the proper version of the Spring portfolio, along with some of the most popular third-party libraries, will already be selected.

There's no need to deal with micromanaging dependency versions. Just bump up the version of Spring Boot and pick up all the improvements.

The Spring Boot team not only releases the software themselves. They also release a **Maven bill of materials** (**BOM**). This is a separate module known as `Spring Boot Dependencies`. Don't panic! It's baked into the modules that are picked up when you adopt Spring Boot.

And with that in place, you can easily pick up new features, bug patches, and any resolved security issues.

> Important
>
> It doesn't matter if you're using *Maven* or *Gradle*. Either build system can consume Spring Boot dependencies and apply their collection of managed dependencies.
>
> We won't go into how Spring Boot dependencies are configured in the build system. Just understand that you can choose the build system you prefer. How to apply this will be covered at the beginning of *Chapter 2, Creating a Web Application with Spring Boot*.

That last part is key, so I'll repeat it: when **Common Vulnerabilities and Exposures** (**CVE**) security vulnerabilities are reported to the Spring team, no matter which component of the Spring portfolio is impacted, the Spring Boot team will make a security-based patch release.

This BOM is released alongside Spring Boot's actual code. All we have to do is adjust the version of Spring Boot in our build file, and everything will follow.

To paraphrase Phil Webb, project lead for Spring Boot, if Spring Framework were a collection of ingredients, then Spring Boot would be a pre-baked cake.

Summary

In this chapter, we discovered the magic of Spring Boot and how it not only brings in Spring beans but also backs off in light of user code. We found out how Spring Boot starters make it easy to add various features of the Spring portfolio as well as some third-party libraries with some simple dependencies. We saw how Spring Boot leverages properties files, allowing us to override various settings of autoconfiguration. We also saw that we can even create our own properties. We learned that Spring Boot manages an entire suite of library dependencies, allowing us to delegate everything to the version of Spring Boot. We also saw how to override that on a one-off basis.

In the next chapter, we'll discover how to apply the concepts from this chapter by building our very first Spring Boot 3 application, starting with the web layer. We'll craft templates and JSON-based APIs, and even stir in a little JavaScript.

Part 2: Creating an Application with Spring Boot

Spring Boot makes it possible to get to the heart of the coding features your users need and not waste time on coding up infrastructure. Instead, you will learn how to serve web templates and JSON APIs. You will then see how to link your web layer to a rich set of database operations. You will also learn how to secure the whole thing so only the right users can access its various features, and finally, how to apply various testing tactics that will evoke confidence in both your team and your consumers.

This part includes the following chapters:

- *Chapter 2, Creating a Web Application with Spring Boot*
- *Chapter 3, Querying for Data with Spring Boot*
- *Chapter 4, Securing an Application with Spring Boot*
- *Chapter 5, Testing with Spring Boot*

2

Creating a Web Application with Spring Boot

In *Chapter 1*, *Core Features of Spring Boot*, we learned about how Spring Boot comes with several powerful features including, autoconfiguration, starters, and configuration properties. Combined with managed dependencies, it becomes easy to upgrade to supported versions of Spring portfolio components as well as third-party libraries.

With the help of **start.spring.io**, we'll learn the basics of creating a web application using Spring Boot. This is vital because the following chapters of this book will build upon this foundation. Since the majority of application development today is focused on web applications, learning how Spring Boot streamlines the whole process will open the door to building apps for years to come!

In this chapter, we'll cover the following topics:

- Using `start.spring.io` to build apps
- Creating a Spring MVC web controller
- Using `start.spring.io` to augment an existing project
- Leveraging templates to create content
- Creating JSON-based APIs
- Hooking in Node.js to a Spring Boot web app

> **Where to find this chapter's code**
> The code for this chapter is available at `https://github.com/PacktPublishing/Learning-Spring-Boot-3.0/tree/main/ch2`.

Using start.spring.io to build apps

The world is littered with different web stacks and toolkits for building web applications and they all come with hooks and modules to tie into various build systems.

But none had the trend-setting notion to help us put together a barebones application directly.

In the past, before Spring Boot entered the scene, we would do one of the following actions to start off a new project:

- *Option 1*: Comb through `stackoverflow.com`, looking for a sample Maven build file
- *Option 2*: Dig through reference documentation, piecing together fragments of build XML, hoping they would work
- *Option 3*: Search various blog sites authored by renowned experts, praying one of their articles contains build details

Oftentimes, we had to contend with out-of-date modules. We may have attempted to apply a configuration option that no longer existed or did not do whatever we needed it to do.

As part of the emergence of Spring Boot came a related website (maintained by the Spring team): **Spring Initializr (start.spring.io)**.

`start.spring.io` comes with the following key features:

- We can select the version of Spring Boot we wish to use
- We can choose our preferred build tool (Maven or Gradle)
- We can enter our project's coordinates (artifact, group, description, and so on)
- We can select which version of Java our project will be built on
- We can choose the various modules (Spring and third-party) to use in our project

We will start by selecting the build tool, language, and version of Spring Boot we wish to use:

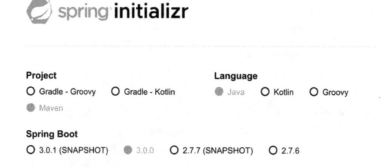

Figure 2.1 – Selecting a build system (Maven), language (Java), and version of Spring Boot

In the preceding screenshot, it's important to point out that we have multiple options. We can choose **Maven** or **Gradle** for our build system and choose between three languages: **Java**, **Kotlin**, or **Groovy**. For this book, we'll stick with Java. But if you or your team wants to leverage the power of Kotlin, you can choose that and get all the right plugins wired into your project to do that.

Spring Initializr also lets us select which version of Spring Boot we wish to use. It's important to point out, that there are subtle differences in the project based on which version we choose, but for this book, we will select 3.0.0. But we don't have to worry about any of that.

It's also nice to know that the website dynamically updates itself when new versions of Spring Boot are released!

After selecting our build system, language, and version of Spring Boot, we need to dial in our project details further down on the page:

Project Metadata

Group	com.springbootlearning.learningspringboot3
Artifact	ch2
Name	Chapter 2
Description	Creating a Web Application with Spring Boot
Package name	com.springbootlearning.learningspringboot3
Packaging	◉ Jar ○ War
Java	◉ 17 ○ 11 ○ 8

Figure 2.2 – Selecting the project coordinates, packaging (JAR), and version of Java (17)

Unless there is some critical reason to select the **War** files (supporting a specific application server or some other legacy reason), it's best to choose the **Jar** files as the packaging mechanism of choice.

"Make JAR not WAR"

– Josh Long, also known as @starbuxman

Why?

WAR files are specific to application servers. Unless you are leveraging one, there is little benefit in using them. JAR files have first-class support from the Spring Boot team, and as we'll see further on in this chapter, they have some keen advantages.

We are also choosing Java 17 for this book. Java 17 is the minimum version required for *Spring Framework 6*, the version underpinning *Spring Boot 3*. In fact, Java 17 is the default option chosen by Spring Initialzr.

The core feature that makes Spring Initializr so nice is that it gives us the ability to select all the modules we wish to include in our project. To do this, go to the **Dependencies** section.

Click on **ADD DEPENDENCIES**; we should be presented with a filter box. Enter web and notice how **Spring Web** rises to the top of the list.

web| Press ⌘ for multiple adds

Spring Web   **WEB**
Build web, including RESTful, applications using Spring MVC. Uses Apache Tomcat as the default embedded   ↵
container.

Figure 2.3 – Adding Spring Web to the project

Hit the *Return* key and see it nicely added to our list.

Amazingly, this is enough to get underway with creating **web controllers**!

Click on the **GENERATE** button at the bottom of the screen.

GENERATE

Figure 2.4 – The GENERATE button, which generates a complete project

Clicking on the preceding button will prompt the download of a ZIP file containing an empty project with a build file containing all our settings for our project.

Unzip the project's ZIP file and open it up inside our favorite IDE, and we can start writing some web controllers!

Tip

It doesn't *really* matter what IDE you use. As mentioned at the beginning of *Chapter 1, Core Features of Spring Boot*, IntelliJ IDEA, Microsoft's VS Code, and Spring Tool Suite all come with support for Spring Boot. Whether it's baked in by default or through the installation of a plugin, you can engage with Spring Boot projects with ease.

With our pre-baked application ready to go, we'll take the first steps toward building a web application in the next section.

Creating a Spring MVC web controller

Assuming we have unzipped that ZIP file from the Spring Initializr and imported it into our IDE, we can immediately start writing a web controller.

But for starters, what is a web controller?

Web controllers are bits of code that respond to HTTP requests. These can comprise an HTTP GET / request that is asking for the root URL. Most websites respond with some HTML. But web controllers can also answer requests for APIs that yield **JavaScript Object Notation (JSON)**, such as HTTP GET /api/videos. Furthermore, web controllers do the heavy lifting of transporting provided JSON when the user is affecting change with an HTTP POST.

The piece of the Spring portfolio that affords us the ability to write web controllers is Spring MVC. Spring MVC is Spring Framework's module that lets us build web apps on top of servlet-based containers using the **Model-View-Controller (MVC)** paradigm.

Yes, the application we are building is Spring Boot. But having picked Spring Web in the previous section, we will put the tried and true **Spring MVC** on our classpath.

In fact, if we peek at the pom.xml file at the root of the project, we'll find one critical dependency:

```
<dependency>
  <groupId>org.springframework.boot</groupId>
  <artifactId>spring-boot-starter-web</artifactId>
</dependency>
```

This is one of those starters mentioned in the previous chapter, in the *Adding portfolio components with Spring Boot starters* section. This dependency puts Spring MVC on our project's classpath. This gives us access to Spring MVC's annotations and other components, allowing us to define web controllers. Its mere presence will trigger Spring Boot's autoconfiguration settings to activate any web controller we create.

There are some other goodies as well, which we'll explore later in this chapter.

Before creating a new controller, it's important to notice that the project already has a base package created per our settings: com.springbootlearning.learningspringboot3.

Let's kick things off by creating a new class inside this package and call it `HomeController`. From there, write the following code:

```
@Controller
public class HomeController {
  @GetMapping("/")
  public String index() {
  return "index";
  }
}
```

This code can be described as follows:

- `@Controller`: Spring MVC's annotation to communicate that this class is a web controller. When the application starts, Spring Boot will automatically detect this class through **component scanning** and will create an instance.

- `@GetMapping`: Spring MVC's annotation to map HTTP `GET` / calls to this method.

- `index`: Because we used the `@Controller` annotation, `index` turns out to be the name of the template we wish to render.

The name of the class and the name of the method aren't critical. They can really be anything. The critical parts are the annotations. `@Controller` signals this class is a web controller, and `@GetMapping` indicates that the `GET` / calls are to be routed to this method.

> **Tip**
> It's good to use class and method names that provide semantic value to us so we can maintain things. In this respect, this (the preceding code snippet) is the controller for the home path of the site we are building.

We mentioned that `index` is the name of the template to render. However, do you remember picking a template engine? That's right, we didn't. In the next section, we'll see how to add a templating engine to our application and use it to start building HTML content.

Using start.spring.io to augment an existing project

What if we *already* started a project and have been working hard on it for the past six months? Creating a brand-new project makes no sense, right?

So, what can we do?

It's possible to pick up an already existing project and make alterations using `start.spring.io`.

We started this chapter with nothing but Spring Web. While we could get pretty far with this, it's not quite enough. While we can write HTML by hand, in this day and age it's easier to use template engines to do that for us. Since we are looking for something lightweight, let's pick **Mustache** (`mustache.github.io`).

If this is appearing a tad contrived, that is because it is. If you're starting a new web project, it makes sense to pick a templating engine *at the same time* that you choose Spring Web. Nevertheless, this tactic of adding additional modules to existing projects still works.

The best way to augment an existing project is to revisit the Spring Initializr site and punch in all our various settings as well as pick the modules we need (especially the new ones we wish to add to our existing project).

Assuming we entered in the same settings we did earlier in this chapter, we just need to click the **DEPENDENCIES** button and enter `mustache`, as shown in the following screenshot:

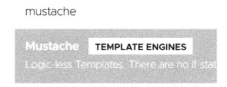

Figure 2.5 – Adding Mustache, a logic-less template language

Hit the *Return* key and add it to the list.

The secret to updating our in-flight project is clicking on the **EXPLORE** button at the bottom of the page instead of the **GENERATE** button we hit earlier.

Figure 2.6 – Exploring a Spring Boot project on the website

The **EXPLORE** button, instead of downloading a ZIP file, lets us view the project we would have gotten right in the browser.

A common tactic is to view the build file: `pom.xml` in our case. From there, we can copy the fragments we need (or copy and paste the whole thing) and paste them into our existing project.

This makes it easy to ensure our project is up to date with any dependencies, customized modules, or whatever.

In this case, we can find the entry for **Mustache**, as shown here:

```
<dependency>
    <groupId>org.springframework.boot</groupId>
    <artifactId>spring-boot-starter-mustache</artifactId>
</dependency>
```

Another Spring Boot starter to the rescue!

Thankfully, this technique helps us reach the Spring Boot starters without having to dig into the details of looking up the names of starters.

The point is, we can quickly get underway with new projects. We can also go back again and again, adding new modules as needed with complete confidence that we aren't breaking things.

Now that we have both Spring Web and Mustache baked into our project, it's time to start creating some real web content, which we'll do in the next section.

Leveraging templates to create content

We can now switch to writing a Mustache template!

After creating the controller class earlier in this chapter, we don't have to do much more. Spring Boot's component scanning feature, as mentioned in *Chapter 1*, *Core Features of Spring Boot*, will do all the legwork of instantiating our controller class. Spring Boot's **autoconfiguration** will add the extra beans that power Mustache's templating engine, hooking it into Spring's infrastructure.

We just need to craft the content that goes inside the template.

By default, Spring Boot expects all templates to be located at `src/main/resources/templates`.

> **Tip**
>
> Spring Boot has **configuration properties** for template engines that default to putting all templates inside `src/main/resources/templates`. Also, each templating engine has a suffix. For Mustache, it's `.mustache`. When we return `index` from a controller method, Spring Boot transforms it into `src/main/resources/templates/index.mustache`, fetches the file, and then pipes it into the Mustache templating engine. It's possible to go in and adjust these settings. But it's frankly easier to just follow the convention.

Create `index.mustache` inside `src/main/resources/templates`. Then, add the following code:

```
<h1>Greetings Learning Spring Boot 3.0 fans!</h1>
```

```
<p>
   In this chapter, we are learning how to make
   a web app using Spring Boot 3.0
</p>
```

This is 100%, bona fide, Use It Anywhere™ HTML5.

To see it in action, we just need to run our application. That's right, we have a fully armed and operational application *already in place*.

Inside our IDE, we just need to right-click on the `Chapter2Application` class that the Spring Initializr created for us and select **Run**.

Once it's up, we can surf over to `localhost:8080` in our favorite browser and see the results:

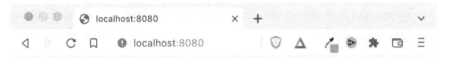

Greetings Learning Spring Boot 3.0 fans!

In this chapter, we are learning how to make a web app using Spring Boot 3.0

Figure 2.7 – A Mustache template rendered with Spring Boot 3

Ta-dah!

What's that? Not impressed?

Okay, we haven't hung any dynamic content here. To be honest, it's boring. A header and a paragraph. Who wants that? Perhaps we should have started with some demo data. This is not hard. The next section will show you how to do this.

Adding demo data to a template

We can tweak the `HomeController` we just made as follows:

```
@Controller
public class HomeController {
   record Video(String name) {}

   List<Video> videos = List.of(
      new Video("Need HELP with your SPRING BOOT 3
```

```
      App?"),
    new Video("Don't do THIS to your own CODE!"),
    new Video("SECRETS to fix BROKEN CODE!"));

  @GetMapping("/")
  public String index(Model model) {
    model.addAttribute("videos", videos);
    return "index";
  }
}
```

Java 17 provides several nice features that we've used in the preceding bit of code:

- We can define a nice little data `Video` object as a Java 17 record with a single line of code
- We can assemble an immutable collection of `Video` objects using `List.of()`

This makes it super simple to create a batch of test data. As for how this works with the templating engine, continue reading.

> **Tip**
>
> Why do we need to create a Java 17 record to encapsulate a single element of data? Mustache operates on named attributes. We could manually write raw JSON with a name and a value, but it's simpler using a Java 17 record. Plus, it gives us stronger type safety. This `Video` type nicely encapsulates the data for our Mustache template.

In order to pass this data on to our template, we need an object that Spring MVC will understand. A holder where we can place data. To do that, we need to add a `Model` parameter to our `index` method.

Spring MVC has a handful of optional attributes we can add to any web method. `Model` is the type we use if we need to hand off data to the templating engine.

The code shown earlier has an attribute named `videos` and is supplied with `List<Video>`. With that in place, we can enhance `index.mustache` to serve it up to the viewers by adding the following code:

```
<ul>
    {{#videos}}
        <li>{{name}}</li>
    {{/videos}}
</ul>
```

The preceding code fragment (placed below the <p> tag we created earlier) can be explained as follows:

- {{#videos}}: Mustache's directive to grab the videos attribute we supplied to the Model object. Because this is a list of data, Mustache will expand this for every entry in the list. It will iterate over each entry of the List<Video> collection stored in Model and create a separate HTML <li> entry.

- {{name}}: Indicates we want the name field of the data construct. This lines up with our Video type's name field. In other words, for each entry of List<Video>, print out the name field between <li> and </li>.

- {{/videos}}: Indicates the end of the looping fragment.

This chunk of Mustache will yield a single HTML unordered list (<ul>) that contains three list items (<li>), each one with a different Video name entry.

> **Tip**
>
> Mustache works with Java's getters, so if we had a value type with getName(), it would serve via {{name}}. But Java 17 records don't generate getters. Instead, the compiler will generate a classfile with name(). Don't worry, Mustache handles this just fine. Either way, we can use {{name}} in our templates.

If we re-run our application and then visit localhost:8080, we can see this updated template in action:

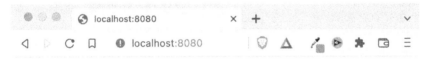

Greetings Learning Spring Boot 3.0 fans!

In this chapter, we are learning how to make a web app using Spring Boot 3.0

- Need HELP with your SPRING BOOT 3 App?
- Don't do THIS to your own CODE!
- SECRETS to fix BROKEN CODE!

Figure 2.8 – Web page with an unordered list of video names

From here, there's no limit to the HTML we can generate. We can even layer in some JavaScript, which is something we'll tackle later in this chapter.

Building our app with a better design

That last app we built was pretty slick. We rapidly modeled some data and served it up in a lightweight template.

The lingering issue is that the design isn't very reusable. The second we need another controller, we'll find ourselves in a tricky situation for the following reasons:

- Controllers shouldn't be managing data definitions. Since they respond to web calls and then interact with other services and systems, these definitions need to be at a lower level.

- Heavyweight web controllers that also deal with data will make it hard to make adjustments as our web needs evolve. That's why it's better if data management is pushed to a lower level.

And so, our first refactoring before continuing forward would be to migrate that `Video` record to its own class, `Video.java`, as follows:

```
record Video(String name) {
}
```

This is the exact same code we wrote earlier, only moved to its own file.

> **Note**
>
> Why isn't the `Video` record marked public? In fact, what visibility does it have? This is Java's *default* visibility, and for classes, records, and interfaces, it defaults to **package-private**. This means that it's only visible to other code in the same package. It's not a bad idea to entertain using Java's default visibility as much as possible and only exposing things outside the package when deemed necessary.

Our next task would be to move that list of `Video` objects into a separate service. Create a class named `VideoService` as follows:

```
@Service
public class VideoService {

  private List<Video> videos = List.of( //
    new Video("Need HELP with your SPRING BOOT 3
      App?"),
    new Video("Don't do THIS to your own CODE!"),
    new Video("SECRETS to fix BROKEN CODE!"));

  public List<Video> getVideos() {
```

```
      return videos;
  }
}
```

`VideoService` can be explained as follows:

- `@Service`: Spring Framework's annotation denoting a class that needs to be picked up during component scanning and added to the application context

- `List.of()`: The same operation used earlier in this chapter to quickly put together a collection of `Video` objects

- `getVideos()`: A utility method to return the current collection of `Video` objects

---

Tip

We have touched on Spring Boot's component scanning functionality briefly in *Chapter 1, Core Features of Spring Boot*, and earlier in this chapter. This is where it shines. We will create a class and then mark it with one of Spring Framework's `@Component`-based annotations, for example, `@Service` or `@Controller`, along with several others. When Spring Boot starts up, one of its first jobs is to run its component scanner, seeking out these classes and instantiating copies. These beans are then registered in the application context, ready to be autowired into any other Spring beans that demand it.

To pivot our `HomeController` to start using this new `VideoService` class, we just need to make updates as follows:

```
@Controller
public class HomeController {

  private final VideoService videoService;

  public HomeController(VideoService videoService) {
    this.videoService = videoService;
  }
}
```

The code we just wrote is quite simple:

- Rip out that `List<Video>` field and replace it with a `private final` instance of `VideoService`

- Populate the new field using **constructor injection**

Let's find out what is **constructor injection** in the next section.

Injecting dependencies through constructor calls

Constructor injection is fancy talk for getting the dependencies a Spring bean needs *through the constructor*. To expand on this, anytime we create a Java class that is picked up by Spring Boot's component scanning feature. Spring Boot will check for any injection points, and if it finds any, will look in the application context for beans of matching type…and inject them!

This is known as **autowiring**. We let Spring handle the issue of finding a Spring bean's dependencies in the application context and plugging them in for us.

Before the days of Spring Boot, autowiring wasn't as popular. Some shops loved it, while others avoided it like the plague. What did those who were opposed do instead? They created a class, marked it up with an `@Configuration` annotation, and created methods using `@Bean` methods. These methods then returned instances of objects to manually wire into other services, either through the constructor or through its setters.

But with the rise of Spring Boot and its beans generated by autoconfiguration that massively leveraged autowiring; autowiring in turn became agreeable to almost everyone.

The following are three ways that we can inject dependencies into a class:

- *Option 1*: The class itself can be marked with one of Spring Framework's `@Component` annotations (or just `@Component` itself) such as `@Service`, `@Controller`, `@RestController`, or `@Configuration`.

- *Option 2*: The `@Autowired` Spring Framework annotation marks points to inject dependencies. It can be applied to constructors, setter methods, and fields (even private ones!).

- *Option 3*: If a class has but one constructor, there is no need to apply the `@Autowired` annotation. Spring will simply assume it's to be autowired.

With `VideoService` injected into `HomeController`, we can update the `index()` method as follows:

```
@GetMapping("/")
public String index(Model model) {
  model.addAttribute("videos", videoService.getVideos());
  return "index";
}
```

The only change in this method from what we coded earlier in this chapter is invoking `videoService` in order to get a list of `Video` objects.

Maybe it felt a bit tedious to make these adjustments to our application, but they will pay off as we continue to flesh things out.

Changing the data through HTML forms

Our web page wouldn't be very impressive if all it did was display server-side data. To make it more dynamic, it should accept new entries send them to our web controller and then show the updated results.

To do this, again, we go back to our Mustache template and start coding a standard HTML form as follows:

```
<form action="/new-video" method="post">
    <input type="text" name="name">
    <button type="submit">Submit</button>
</form>
```

This simple addition to `index.mustache` can be explained as follows:

- This HTML form will result in a `POST /new-video` call to our server-side app.
- It has a single text-based input called `name`.
- This form takes effect when the user clicks on the `Submit` button.

If you're wondering where all the Mustache stuff is, there is none. An HTML form is pretty simple. It's possible to render dynamic content here if needed, but our focus is to get the submission of new data working with a relatively simple scenario.

To make our Spring Boot application respond to `POST /new-video`, we need to write another controller method to our `HomeController` as follows:

```
@PostMapping("/new-video")
public String newVideo(@ModelAttribute Video newVideo) {
  videoService.create(newVideo);
  return "redirect:/";
}
```

This additional web controller method can be explained as follows:

- `@PostMapping("/new-video")`: Spring MVC's annotation to capture `POST` `/new-video` calls and route them to this method.
- `@ModelAttribute`: Spring MVC's annotation to parse an incoming HTML form and unpack it into a `Video` object.
- `videoService.create()`: Not-yet-written method for storing the new video object.

- "redirect:/": Spring MVC directive that sends the browser an **HTTP 302 Found** to URL /. A 302 redirect is the standard for a soft redirect. (301 is a permanent redirect, instructing the browser to not try the original path again.)

This additional web controller method now demands that we augment our VideoService with the means to add more Video objects.

It's important to recognize that so far, we've been using Java 17's List.of() operator to build our collection of videos, which produces an immutable list. This immutable list honors Java's List interface, giving us access to an add() method. If we try to use it, it will only generate an UnsupportedOperationException.

No, we need to take a couple of extra steps if we are going to mutate this immutable collection.

The recipe for adding to an immutable *anything* is to create a new immutable instance out of the original contents combined with new contents. This is where we can take advantage of more familiar List-based APIs:

```
public Video create(Video newVideo) {
    List<Video> extend = new ArrayList<>(videos);
    extend.add(newVideo);
    this.videos = List.copyOf(extend);
    return newVideo;
}
```

This addition to our VideoService can be explained as follows:

- The signature of the method calls for a new video and then returns that same one back (common behavior for a repository-style service).

- new ArrayList<>(): creates a new ArrayList, a mutable collection, using its List-based constructor. This new collection initializes with the proper size and then copies every entry into the new ArrayList.

- This ArrayList does have a usable add() method that allows us to add our new Video object onto the end.

- Java 17 provides a copyOf() operator that takes any existing List and copies all of its elements into a new immutable list.

- Finally, we return the new Video object.

It's important to point out that while it took us a couple of extra steps, the preceding code ensures that no existing copy of data will be accidentally mutated by invoking a method. Side effects are prevented by doing this, ensuring a consistent state is maintained.

> **Note**
>
> While this data may be consistent thanks to using immutable lists, it is by no means thread-safe. If multiple POST calls were made to the endpoint we just defined, they would all attempt to update the same VideoService, probably resulting in some form of a race condition that could cause loss of data. Given the fact that there are entire books written about solving such problems, we aren't going to focus on making our code bulletproof in this one.

With these changes in place, we can re-run our application and check out our enhanced UI:

Greetings Learning Spring Boot 3.0 fans!

In this chapter, we are learning how to make a web app using Spring Boot 3.0

- Need HELP with your SPRING BOOT 3 App?
- Don't do THIS to your own CODE!
- SECRETS to fix BROKEN CODE!
- Learning Spring Boot 3
  [   ] Submit

Figure 2.9 – Mustache template with an HTML form

If we were to enter Learning Spring Boot 3 in the input and click **Submit**, the controller will issue a redirect back to /. The browser will navigate back to the root path, causing it to fetch the data and render it with our latest Video.

> **Further reading**
>
> Perhaps you wish to learn more about Mustache and how it interacts with Spring Boot? Check out Dave Syer's article *The Joy of Mustache: Server Side Templates for the JVM* (https:// springbootlearning.com/mustache) where he goes into detail about integrating Spring Boot with Mustache, including ways to have a consistent layout and even code custom Mustache lambda functions.

With that, we have a functional web page that serves up dynamic content that also lets us add more content. But this is by no means complete. We should investigate adding support for building web-based APIs in the next section.

Creating JSON-based APIs

A key ingredient in building any web application is the ability to provide an **API**. In the olden days, this was complex and hard to ensure compatibility.

In this day and age, the world has mostly converged on a handful of formats, many based on **JSON**-based structures.

One of the powerful features of Spring Boot is that when you add Spring Web to a project, as we did at the beginning of this chapter, it adds **Jackson** to the `classpath`. Jackson is a JSON serialization/deserialization library that has been widely adopted by the Java community.

Jackson's ability to let us define how to translate Java classes back and forth with our preferred flavor of JSON combined with Spring Boot's ability to autoconfigure things means that we don't have to lift another finger of setup to start coding an API.

To start things off, we create a new class in the same package we've been using throughout this chapter. Call it `ApiController`. At the top, apply the `@RestController` annotation.

`@RestController` is similar to the `@Controller` annotation we used earlier. It signals to Spring Boot that this class should be automatically picked up for component scanning as a Spring bean. This bean will be registered in the application context and also with Spring MVC as a controller class so it can route web calls.

But it has one additional property—it switches every web method from being template-based to JSON-based. In other words, instead of a web method returning the name of a template that Spring MVC renders through a templating engine, it instead serializes the results using Jackson.

Look at the following code:

```
@RestController
public class ApiController {

  private final VideoService videoService;

  public ApiController(VideoService videoService) {
    this.videoService = videoService;
  }

  @GetMapping("/api/videos")
  public List<Video> all() {
    return videoService.getVideos();
  }
}
```

We can describe the preceding code in detail as follows:

- We already mentioned how the `@RestController` annotation marks this as a Spring MVC controller that returns JSON

- Using constructor injection, we will automatically get a copy of the same `VideoService` we created earlier in this chapter

- `@GetMapping` responds to HTTP GET calls from `/api/videos`

- This web method will fetch that list of `Video` records and return them, causing them to get rendered into a JSON array by Jackson

In fact, if we run the application right now and curl that endpoint, we can see the following results:

```
[
    {
        "name": "Need HELP with your SPRING BOOT 3 App?"
    },
    {
        "name": "Don't do THIS to your own CODE!"
    },
    {
        "name": "SECRETS to fix BROKEN CODE!"
    }
]
```

This barebones JSON array contains three entries, one for each of our `Video` records. And since the `Video` record only has one attribute, `name`, this is exactly what Jackson yields.

There is no need to configure *anything* to make Jackson start producing.

Of course, an API that does nothing but *produce* JSON isn't much of an API at all. Naturally, we need to *consume* JSON as well.

To do that, we need to create a web method in our `ApiController` class that will respond to HTTP POST calls.

POST versus PUT versus everything else

There are several standard HTTP verbs at our disposal. The most common ones are GET, POST, PUT, and DELETE. There are actually others, but we won't be needing them here. It's important to understand that GET calls are expected to do nothing but return data. They are not meant to cause state changes on the server, which is also known as idempotent. POST calls are what are used to introduce new data to the system. This is analogous to inserting a new row of data into a relational database table.

PUT is similar to POST in that it's used for making changes, but it's better described as updating existing records. It's possible to also *update* a non-existent record, but this depends on how things have been set up on the server. Finally, DELETE is used to remove something from the server.

It is somewhat common behavior, although not a required standard, that any updates to the system should return to the person or system making the request, a copy of the new or deleted entry.

Add the following code to the ApiController class, right below the all() method:

```
@PostMapping("/api/videos")
public Video newVideo(@RequestBody Video newVideo) {
    return videoService.create(newVideo);
}
```

The preceding code can be described as follows:

- @PostMapping: Maps HTTP POST calls to /api/videos onto this method

- @RequestBody: Spring MVC's annotation to signal that the incoming HTTP request body should be deserialized via Jackson into the newVideo argument as a Video record

- We then delegate the actual handling of this incoming Video record to our VideoService, returning back the record after it's been added to the system

We already coded this create() operation earlier in this chapter, so there's no need to go back into it.

Tip

Earlier in the section, we said "curl that endpoint" and saw a chunk of JSON printed out. **curl** (https://curl.se/) is a popular command-line tool that lets you interact with web APIs. In fact, this summary probably doesn't do the tool justice. Suffice it to say, you probably want to install it on your system.

With our new addition to our API controller, we can interact with it from the command line as follows:

```
$ curl -v -X POST localhost:8080/api/videos -d '{"name":
"Learning Spring Boot 3"}' -H 'Content-type:application/json'
```

The preceding command can be described as follows:

- `-v`: Asks for curl to produce verbose output, providing extensive details about the whole interaction.

- `-X POST`: Signals it to use `HTTP POST` instead of the default `GET` call.

- `localhost:8080/api/video`: Provides the URL to direct the command.

- `-d '{…}'`: Provides the data. Since the fields in JSON are delimited with double quotes, the entire JSON document is handed to curl using single quotes.

- `-H 'Content-type:application/json'`: Provides the HTTP header that alerts the web app that this is a JSON-formatted request body.

The results of the command are as follows:

```
* Connected to localhost (::1) port 8080 (#0)
> POST /api/videos HTTP/1.1
> Host: localhost:8080
> User-Agent: curl/7.64.1
> Accept: */*
> Content-type:application/json
> Content-Length: 34
>
* upload completely sent off: 34 out of 34 bytes
< HTTP/1.1 200
< Content-Type: application/json
{"name":"Learning Spring Boot 3"}
```

The preceding output response shows the following:

- The command is shown toward the top including the HTTP verb, the URL, and the headers (`User-Agent`, `Accept`, `Content-type`, and `Content-Length`)

- The response is shown toward the bottom including an HTTP `200` success status code

- The response is denoted as `application/json`

- The actual response body contains the JSON-formatted new `Video` entry we created

Our method worked!

We can further verify this by re-pinging the /api/videos endpoint and looking for our latest addition:

```
$ curl localhost:8080/api/videos
[
    {"name":"Need HELP with your SPRING BOOT 3 App?"},
    {"name":"Don't do THIS to your own CODE!"},
    {"name":"SECRETS to fix BROKEN CODE!"},
    {"name":"Learning Spring Boot 3"}
]
```

The preceding code shows our latest entry at the bottom.

What do we have now? We have a web application with two key facets: a template-based version rendered in the browser for humans to read and a JSON-based version consumable by third parties.

With all this in place, we are finally able to write a little JavaScript!

Hooking in Node.js to a Spring Boot web app

So, do we need our web app to use JavaScript? To be honest, what web app *doesn't* need JavaScript? It's only the de facto standard tool found in every web browser on the planet.

In case you didn't know it, JavaScript is a completely different world when it comes to tools and app building. So, how do we cross this vast gulf between Java and JavaScript developer tooling?

In a nutshell, we need to enter the world of Node.js. And to our fortune, there is a Maven plugin that can bridge this gap for us, known as the Maven frontend plugin (**frontend-maven-plugin**).

This plugin unites Node.js actions with Maven's lifecycles, allowing us to properly invoke Node.js at the right time to download packages and assemble JavaScript code into a bundle.

Of course, compiling and bundling a JavaScript payload is for naught if there was no way for Spring Boot to bring it online.

Thankfully, Spring Boot has a solution. Anything found in src/main/resources/static is automatically picked up and will be put on the base path of our web application when fully assembled. This means we simply need to direct our Node.js bundling tool to drop its final results *there*.

If this is starting to sound confusing…well it is. Or at least it can be. So, let's take things one step at a time, starting with the frontend-maven-plugin. If we open up the pom.xml file created by start.spring.io, there should be an entry about two-thirds of the way down called <plugins>. It should already have an entry for spring-boot-maven-plugin.

Right below `spring-boot-maven-plugin`, add another `<plugin>` entry as follows:

```
<plugin>
  <groupId>com.github.eirslett</groupId>
  <artifactId>frontend-maven-plugin</artifactId>
  <version>1.12.1</version>
  <executions>
    <execution>
      <goals>
          <goal>install-node-and-npm</goal>
      </goals>
    </execution>
  </executions>
  <configuration>
    <nodeVersion>v16.14.2</nodeVersion>
  </configuration>
</plugin>
```

This addition to our `pom.xml` build file can be explained as follows:

- We've added the coordinates for the latest version of `frontend-maven-plugin` (at the time of writing)

- Right now, it has one execution, `install-node-and-npm`. This command will download **Node.js** and its package manager **npm**

- In the configuration section toward the bottom, it specifies the latest **long-term support (LTS)** version of Node.js

This plugin does its thing during Maven's `generate-resources` phase. You can see it right away with the following output shown in the console:

```
$ ./mvnw generate-resources
[INFO] --- frontend-maven-plugin:1.12.1:install-node-and-npm
(default) @ ch2 ---
[INFO] Installing node version v16.14.2
[INFO] Downloading https://nodejs.org/dist/v16.14.2/node-
v16.14.2-darwin-x64.tar.gz to /Users/gturnquist/.m2/repository/
com/github/eirslett/node/16.14.2/node-16.14.2-darwin-x64.tar.gz
[INFO] Unpacking /Users/gturnquist/.m2/repository/com/github/
eirslett/node/16.14.2/node-16.14.2-darwin-x64.tar.gz into /
Users/gturnquist/src/learning-spring-boot-3rd-edition-code/ch2/
node/tmp
```

```
[INFO] Copying node binary from /Users/gturnquist/src/learning-
spring-boot-3rd-edition-code/ch2/node/tmp/node-v16.14.2-
darwin-x64/bin/node to /Users/gturnquist/src/learning-spring-
boot-3rd-edition-code/ch2/node/node
[INFO] Extracting NPM
[INFO] Installed node locally.
```

It should be pointed out that the frontend-maven-plugin actually downloads and unpacks Node.js, npm, and **node package execute** (**npx**) in our project's root directory underneath the node folder.

> **Tip**
> Node.js and all its tools and modules can be considered intermediate build artifacts. There's no need to commit them to version control. So, be sure to add the node folder as well as the intermediate node_modules folder to your list of items to not commit (for example, add node and node_modules to the project's .gitignore file).

With this plugin in place, we are ready to embrace the world of JavaScript in the next section.

Bundling JavaScript with Node.js

At this point, we have the tooling. But no actual modules.

To start adding modules, we'll use npm. And the first thing we must do is pick a Node.js package bundler. There are many to pick from, but let's pick **Parcel** by typing the following command:

```
% node/npm install --save-dev parcel
```

This will use the locally installed copy of Node.js and its npm command to create a package.json file. The –save-dev option signals that this is a development module, and not a package used by our app.

Now that we have a package.json file created for our project, we need to hook it into the frontend-maven-plugin. To do that, we need to add another <execution> entry as follows:

```
<execution>
  <id>npm install</id>
  <goals>
    <goal>npm</goal>
  </goals>
</execution>
```

This additional fragment will configure the `frontend-maven-plugin` to run `npm install`, the command that will build our JavaScript bundle.

Now so far, we don't have much in packages. Just the Parcel build tool. Before we start adding JavaScript modules, we should probably configure it to build things properly. Edit the `package.json` that npm just created as shown here so that Parcel will assemble an ES6 module for us:

```
{
    ...
    "source": "src/main/javascript/index.js",
    "targets": {
      "default": {
        "distDir": "target/classes/static"
      }
    },
    ...
}
```

This addition to `package.json` will cause the following to happen:

- `Source`: Points at a yet-to-be-written `index.js` JavaScript file. This will be our entry point for our JavaScript app. As far as Parcel is concerned, it doesn't really matter where this file is. Since we're using a Maven-based project, we can use `src/main/javascript`.

- As for the target destination, we can configure the default target with a `distDir` setting of `target/classes/static`. Parcel supports building multiple targets such as different browsers, but we don't need that. A single, default destination will do. By putting the results in the target folder, anytime we run a *Maven clean* cycle, this compiled bundle will be cleaned out.

While npm is Node.js' tool for downloading and installing packages, npx is Node.js' tool for running commands. By adding another `<execution>` entry to the `frontend-maven-plugin`, we can have it run Parcel's `build` command:

```
<execution>
  <id>npx run</id>
  <goals>
    <goal>npx</goal>
  </goals>
  <phase>generate-resources</phase>
  <configuration>
    <arguments>parcel build</arguments>
```

```
    </configuration>
  </execution>
```

This extra step runs `npx parcel build` after the `npm install` command is run, ensuring Parcel does its build step.

With all this, we can start installing some Node packages to build a sophisticated frontend in the next section.

Creating a React.js app

There are so many ways to build JavaScript apps these days and they all have their own features and benefits. For illustrative purposes, we'll use React.js, Facebook's toolkit for app building.

Type the following command:

```
node/npm install --save react react-dom
```

The preceding command will update `package.json` with the `react` and `react-dom` modules. Now, we can start writing some JavaScript!

Create `index.js` inside `src/main/javascript` as follows:

```
import ReactDOM from "react-dom"
import { App } from "./App"
const app = document.getElementById("app")
ReactDOM.render(<App />, app)
```

This is our entry point, as shown earlier in `package.json`:

- The first line imports `ReactDOM`, a key module needed to launch React
- The second line imports the custom UI we'll build further down in this section
- The third line uses vanilla JavaScript to find the element on the web page with `id="app"` where we can "hang" our app
- The fourth line actually renders the `<App/>` component in our soon-to-be-coded app using `ReactDOM`

React operates with a top-down perspective. You render a top-level component, and then that component, in turn, renders components nested further down.

It also uses a shadow **document object model** (**DOM**) where we don't specify the actual nodes to render, but instead use virtual nodes. React computes the changes from whatever the current state is and generates the changes.

To continue building this application, we need to create `App.js` in `src/main/javascript` as follows:

```
import React from 'react'
import ListOfVideos from './ListOfVideos'
import NewVideo from "./NewVideo"

export function App() {
    return (
        <div>
            <ListOfVideos/>
            <NewVideo/>
        </div>
    )
}
```

The preceding chunk of JavaScript has some key parts:

- We import React to build components.

- The local bits of JavaScript that we will write after this includes a list of videos (`ListOfVideos.js`) and `from` for creating new videos (`NewVideo.js`).

  From here, we have a single function, `App()`, exported publicly. Because it returns some HTML-styled elements, this signals to Parcel that we're working with **JavaScript XML (JSX)**, which contains some more React components to render.

> **React and JSX**
>
> React introduces a concept called JSX, where we can combine unique HTML elements with JavaScript code. In the past, we were told that mixing HTML and JavaScript was bad. But in truth, when laying out a UI that ties together functions, JSX provides an excellent blend. Instead of using tricky functions to layer JavaScript on top of HTML, React works by allowing us to build up tiny bits of HTML combined cohesively with functions that support their operation. Combined with its internal state management, React has become quite popular among many for building web apps.

The first thing to replicate from our earlier template is listing all the videos from the backend. To generate the same HTML unordered list in React, create `ListOfVideos.js` in `src/main/javascript` as follows:

```
import React from "react"
```

```
class ListOfVideos extends React.Component {
    constructor(props) {
        super(props)
        this.state = {data: []}
    }

    async componentDidMount() {
        let json = await fetch("/api/videos").json()
        this.setState({data: json})
    }

    render() {
        return (
            <ul>
                {this.state.data.map(item =>
                    <li>
                        {item.name}
                    </li>)}
            </ul>
        )
    }
}

export default ListOfVideos
```

The preceding React component can be described as follows:

- This code uses ES6 classes that extend React.Component.

- Its constructor creates a state field to maintain an internal state.

- componentDidMount() is the function called by React right after this component is inserted into the DOM and rendered. It uses vanilla JavaScript's fetch() function to retrieve data from the JSON API we created earlier in this chapter. Because that function returns a promise, we can use the ES6 await function to wait for results and then update the internal state using React.Component's setState(). For this method to work with the rest of things properly, we must mark it async. It's also important to understand that anytime setState() is invoked, React will re-render the component.

- The sizzle is in the `render()` method where we actually lay out HTML elements (or more React components). This code uses the internal state and maps over the array of data, converting each piece of JSON into an HTML line item. No elements? No line items!

In the preceding chunk of code, we mentioned both properties and state. Properties typically comprise information injected into a React component from the outside. State is maintained within. It's possible to initialize the state from properties, or as shown in this code, the component itself can fetch the data stored in the state.

What's important to clarify, is that properties are typically considered immutable *within the React component that has them injected*. State is meant to evolve and change, in turn driving our rendered elements.

Our React app wouldn't be much if we couldn't create new entries. So, let's create a new component that duplicates the HTML form we created earlier in this chapter in the *Changing the data through HTML forms* section!

Create `NewVideo.js` as shown here in `src/main/javascript`:

```
import React from "react"

class NewVideo extends React.Component {
    constructor(props) {
        super(props)
        this.state = {name: ""}
        this.handleChange = this.handleChange.bind(this);
        this.handleSubmit = this.handleSubmit.bind(this);
    }

    handleChange(event) {
        this.setState({name: event.target.value})
    }

    async handleSubmit(event) {
        event.preventDefault()
        await fetch("/api/videos", {
            method: "POST",
            headers: {
                "Content-type":
                    "application/json"
            },
```

```
                   body: JSON.stringify({name: this.state.name})
            }).then(response =>
               window.location.href = "/react")
       }


      render() {
          return (
             <form onSubmit={this.handleSubmit}>
                 <input type="text"
                         value={this.state.name}
                         onChange={this.handleChange}/>
                 <button type="submit">Submit</button>
             </form>
          )
      }
}


export default NewVideo
```

This React component has some of the same bits as the other one, such as the `import` statement as well as a JavaScript class extending `React.Component`. But it contains some different parts, which are as follows:

- It has the `handleChange` and `handleSubmit` functions, both bound to the component. This ensures that `this` will properly reference the component upon invocation.

- The `handleChange` function is invoked anytime the field on the form is altered. It updates the component's internal state.

- The `handleSubmit` function is invoked when the button is clicked. It disables standard JavaScript *bubble-up* behavior. Instead of button click events trickling up through the stack, it's handled right here, invoking a vanilla JavaScript `fetch()` to affect a POST call on the `/api/videos` endpoint created earlier in the *Creating JSON-based APIs* section of this chapter.

- The `render()` function creates an HTML form element with the `onSubmit()` event tied to the `handleSubmit` function and the `onChange` event tied to the `handleChange` function.

Another aspect of this code is the async/await modifiers used on the `handleSubmit` function. Some JavaScript functions return standard promises (https://promisesaplus.com/), such as its built-in `fetch` function. To ease the usage of these APIs (and save us from using third-party libraries), ES6 introduced the `await` keyword, allowing us to indicate we wish to wait for the results. To support this, we must flag the function itself as `async`.

To load up the React app we've laid out, we need a separate Mustache template. Create `react.mustache` inside `src/main/resources/templates` and include the following elements:

```
<div id="app"></div>
<script type="module" src="index.js"></script>
```

The preceding code contains two critical aspects:

- `<div id="app" />` is the element that the React component `<App />` will be rendered at per the `document.getElementById("app")` from earlier in this section.
- The `<script>` tag will load our app through the `index.js` bundle Parcel will build. The `type="module"` argument indicates it's an ES6 module.

The rest of `react.mustache` can have the same header and paragraph as our other template has.

To serve up our React app, we need a separate web controller method to `HomeController`, as shown here:

```
@GetMapping("/react")
public String react() {
  return "react";
}
```

This will serve the `react` Mustache template when the user requests `GET /react`.

With all this effort, perhaps you're wondering if it was worth it. After all, we simply duplicated the content of a template. It took considerably more effort. If that was all we did, it's true. This *is* too much effort.

But React really kicks in when we need to design a much more complex UI. For example, if we needed various components to optionally render or needed different types of components to appear, all driven by the internal state of things, that's where React begins to shine.

As stated earlier in this section, React also has the shadow DOM. We don't have to focus on the somewhat outdated concept of finding parts of the DOM and manually updating them. Instead, with React, we push out a set of HTML components. Then, as the internal state updates, the components are re-rendered. React simply computes the changes on real DOM elements and automatically updates itself. We don't have to handle that.

But enough about React. The focus of this section was to illustrate how to merge JavaScript into a Spring Boot web application. These techniques of setting up Node.js, installing packages, and utilizing a build tool work, irrespective of whether we're using React, Angular, Vue.js, or whatever.

If we have static components, be they JavaScript or CSS, we can put them into our `src/main/resources/static` folder. If they are generated, such as a compiled and bundled JavaScript module, we've seen how to route that output to `target/classes/static`.

In short, we've managed to connect the powerful world of Node.js and JavaScript with the land of Spring Boot and Java.

Summary

In this chapter, we used `start.spring.io` to create a barebones web application. We injected some demo data using a service. We create a web controller that uses Mustache to render dynamic content based on the demo data.

Then, we created a JSON-based API allowing third-party apps to interact with our web app, whether it's to retrieve data or send in updates.

Finally, we leveraged Node.js to introduce some JavaScript to our web app using a Maven plugin.

Building web controllers, serving up templates, rendering JSON-based APIs, and serving up JavaScript apps is a valuable skill on just about any project.

In the next chapter, *Querying for Data with Spring Boot*, we will dig into creating and managing real data using Spring Data and the amazing power Spring Boot brings us.

3

Querying for Data with Spring Boot

In the previous chapter, we learned how Spring Boot manages embedded servlet containers, automatically registers our web controllers, and even provides JSON serialization/deserialization, easing the creation of APIs.

What application doesn't have data? Spoiler alert – none. That's why this chapter is focused on learning some of the most powerful (and handy) ways to store and retrieve data.

In this chapter, we'll cover the following topics:

- Adding Spring Data to an existing Spring Boot application
- DTOs, entities, and POJOs, oh my!
- Creating a Spring Data repository
- Using custom finders
- Using Query By Example to find tricky answers
- Using the custom **Java Persistence API (JPA)**

Being able to store and retrieve data is a critical need for any application, and this list of topics will provide you with vital abilities.

> **Where to find this chapter's code**
> The source code for this chapter can be found at `https://github.com/PacktPublishing/Learning-Spring-Boot-3.0/tree/main/ch3.`

Adding Spring Data to an existing Spring Boot application

Imagine we have an application brewing. We showed our program manager some preliminary web pages based on the pitch she is hastily putting together. While excited at that, she signals we need to hook them up to some *real* data.

But instead of swallowing with dread, we smile from ear to ear. **Spring Data** is the ticket to powerful data management.

Before we can move forward, though, we must make a choice. What data store exactly do we need?

The most common database used today is a relational one (*Oracle*, *MySQL*, *PostgreSQL*, and so on). As mentioned in a past *SpringOne* keynote, relational databases comprise 80% of all projects created on Spring Initializr. Choosing a **NoSQL** (**not only SQL**) data store requires careful consideration, but here are three options we can explore:

- **Redis** is principally built as a key/value data store. It's very fast and very effective at storing huge amounts of key/value data. On top of that, it has sophisticated statistical analysis, time-to-live, and functions.

- **MongoDB** is a document store. It can store multiple levels of nested documents. It also has the ability to create pipelines that process documents and generate aggregated data.

- **Apache Cassandra** provides a table-like structure but has the ability to control consistency as a trade-off with performance.

SQL data stores have historically had hard requirements on predefining the structure of the data, enforcing keys, and other aspects that tend to give no quarter.

NoSQL data stores tend to relax some of these requirements. Many don't require an upfront schema design. They can have optional attributes, such that one record may have different fields from those of other records (in the same keyspace or document).

Some NoSQL data stores give you more flexibility when it comes to scalability, eventual consistency, and fault tolerance. For example, Apache Cassandra lets you run as many nodes as you like and lets you choose how many nodes have to agree on the answer to your query before giving it. A faster answer is less reliable, but 75% agreement may be faster than waiting for all nodes (as is typically the case with relational data stores).

NoSQL data stores typically do *not* support transactions. Some of them are starting to offer them, in limited contexts. But in general, if a NoSQL data store were to mimic every feature of relational data stores (consistency, transactions, and fixed structure), they'd probably lose the features that make them faster and more scalable.

With that being said, let's focus on using a traditional data store. The features described in the rest of this chapter are widely available with *any* data store supported by Spring Data. The next section will explain why.

Using Spring Data to easily manage data

Spring Data has a unique approach to simplifying data access. Spring Data doesn't use the *lowest common denominator* approach. This is the tactic where a single API is defined in an interface, and an implementation is offered for every data store.

This tends to reduce access to *only* the features that *all* data stores share. Since all data stores offer varying features, the keen aspects of a given data store that make us want to use it usually aren't found in that shared API!

No, Spring Data takes a different approach. Each data store has multiple ways to access data, but *not* with an identical API. For starters, almost every Spring Data module has a **template** that gives us easy access to data store-specific features. Some of these templates include the following:

- `RedisTemplate`

- `CassandraTemplate`

- `MongoTemplate`

- `CouchbaseTemplate`

These template classes aren't descendants of some common API. Each Spring Data module has its own core template. They all have a similar lifecycle where they handle resource management. Each template has a mixture of functions, some based on a common data access paradigm and others based on the data store's native features.

While we can do just about anything using a data store's template, there are easier ways to access data. Many Spring Data modules include **repository** support, making it possible to define queries, updates, and deletes, based purely on our domain types. We'll see how to use more of this later in the chapter.

There are additional ways to define data requirements, including Query By Example and support for the third-party library **Querydsl**.

> **Note**
>
> While every data store has a template, `HibernateTemplate`, a long-time part of Spring Framework's Hibernate solution, is really a tool meant to help legacy apps migrate to Hibernate's `SessionFactory.getCurrentSession()` API. The Hibernate team prefers using this approach or migrating toward using JPA's `EntityManager` directly. For this reason, we won't be delving into `HibernateTemplate` in this chapter. However, we will explore the many ways that Spring Data JPA simplifies access to relational data stores.

There's a common undercurrent in all these approaches. Writing the `select` statements, whether they are for Redis, MongoDB, or Cassandra, is not only tedious but also costly to maintain. Considering a huge portion of queries are simply copying in structural values that map onto domain types and field names, Spring Data leverages domain information to help developers write queries.

Instead of rolling queries by hand, shifting the language of data access to domain objects and their fields lets us shift our problem-solving to business use cases.

Nevertheless, there is always a handful of use cases that require handwritten queries. For example, there are monthly customer deliverables that require joining 20 tables or a customer report that has a varying mix of inputs.

There is always the option to sidestep any of Spring Data's help and instead write the query directly.

Through the rest of this chapter, we'll explore how these various forms of data access provided by Spring Data let us focus on solving customer problems instead of battling query typos.

Adding Spring Data JPA to our project

Before we can do anything with Spring Data, we must add it to our project. While we spent a bit of time in the previous section discussing various data stores, let's settle on a relational database.

To do that, we will use Spring Data JPA. To get off the ground, we'll choose a simple embedded database, **H2**. This database is a **Java Database Connectivity (JDBC)** based relational database written in Java. It's effective for prototyping efforts.

To add **Spring Data JPA** and **H2** to our already drafted app, we can easily use the same tactic from *Chapter 2, Creating a Web Application with Spring Boot* (using `start.spring.io` to build apps):

1. Visit `start.spring.io`.
2. Enter the same project artifact details as before.
3. Click on **DEPENDENCIES**.
4. Select **Spring Data JPA** and **H2**.
5. Click on **EXPLORE**.
6. Look for the `pom.xml` file and click on it.
7. Copy the new bits onto the clipboard.
8. Open up our previous project inside our IDE.
9. Open our `pom.xml` file and paste the new bits into the right places.

Hit the refresh button in the IDE, and we're ready to go!

Now with Spring Data JPA and H2 added to our project, we are ready to start designing our data structures in the next section!

DTOs, entities, and POJOs, oh my!

Before we start slinging code, we need to understand a fundamental paradigm: **data transfer objects (DTOs)** versus **entities** versus **plain old Java objects (POJOs)**.

The differences between these three conventions aren't something that is directly enforced by any sort of tool. That's why it's a paradigm and not a coding construct. So, what exactly are DTOs, entities, and POJOs?

- **DTO**: A class whose purpose is to transfer data, usually from server to client (or vice versa)

- **Entity**: A class whose purpose is to store/retrieve data to/from a data store

- **POJO**: A class that doesn't extend any framework code nor has any sort of restrictions baked into it

Entities

When we write code to query data from a database, the class where our data ends up is commonly called an **entity**. This concept was turned into a standard when JPA was rolled out. Literally, every class involved with storing and retrieving data through JPA must be annotated with @Entity.

But the concept of entities doesn't stop with JPA. Classes used to ferry data in and out of MongoDB, although they require no such annotation, can also be considered entities. Classes that are involved in data access typically have requirements levied on them by the data store. JPA specifically wraps entity objects returned from queries with proxies. This lets the JPA provider track updates so that it knows when to actually push updates out to the data store (known as **flushing**) and also helps it to better handle entity caching.

Speaking of entities, we need to lay out the details for the video type we wish to store in the database in this chapter. The following code will be suitable for this chapter's needs:

```
@Entity
class VideoEntity {

  private @Id @GeneratedValue Long id;
  private String name;
  private String description;

  protected VideoEntity() {
    this(null, null);
  }
```

```
    VideoEntity(String name, String description) {
      this.id = null;
      this.description = description;
      this.name = name;
    }
  // getters and setters
}
```

This preceding code has the following characteristics:

- `@Entity` is JPA's annotation to signal this is a JPA-managed type.

- `@Id` is JPA's annotation to flag the primary key.

- `@GeneratedValue` is a JPA annotation to offload key generation to the JPA provider.

- JPA requires a no-argument constructor method that is either public or protected.

- We also have a constructor where the `id` field isn't provided: a constructor designed for creating new entries in the database. When the `id` field is null, it tells JPA we want to create a new row in the table.

We aren't going to spend a lot of time talking about modeling entities with JPA. There are entire books dedicated to the intricacies of entity modeling.

DTOs

DTOs, on the other hand, are typically used in the web layer of applications. These classes are more concerned with taking data and ensuring it's properly formatted for either HTML generation or JSON handling. **Jackson**, Spring Boot's default JSON serialization/deserialization library, comes with a fistful of annotations to customize JSON rendering.

> **Note**
> It should be said that DTOs aren't confined to JSON. Using XML or any other form of data exchange format has the same need of ensuring proper formatting of data. JSON just happens to be the most popular format in today's industry; hence, the reason Spring Boot puts Jackson on `classpath` by default when we choose Spring Web.

Why do we need DTOs and entities? Because a keen lesson learned over recent years is that classes are easier to maintain and update if they try to concentrate on doing just one task. In fact, there's a name for this concept: the **single-responsibility principle** (**SRP**).

A class that tries to be both a DTO and an entity is harder to manage in the long run. Why? Because there are two stakeholders: the web layer and the persistence layer.

> **Tip**
>
> Short-term versus long-term goals. Notice how I said that a class that tries to be both a DTO and an entity is harder to manage in the long run? That's true. But what about that demo you give to your CTO when you're pitching Spring Boot? That is a short-term scenario where you need to get a point across. You aren't trying to build a long-lasting app, but instead, a quick demo. In those scenarios, it's okay to let one class serve as both DTO and entity. But anything slated for production will probably be better maintained over the long run if those two ideas are decoupled. For a more detailed discussion, check out my video, *DTOs vs. Entities*, at `https://springbootlearning.com/dtos-vs-entities`.

We've mentioned DTOs and entities. Where do POJOs fit into all this?

POJOs

Spring has a long history of supporting a POJO-oriented programming style. Before Spring, many if not most Java-based frameworks required developers to extend various classes. This drives user code to hook into the framework to make things happen.

These sorts of classes were hard to work with. They didn't lend themselves to writing test cases due to requirements inherited from the framework. It often required spinning everything up to verify that user-created code was working right. Overall, it made for a heavy-handed coding experience.

A POJO-based approach meant writing user code that *didn't* have to extend any framework code.

Spring's concept of registering beans with an application context made it possible to avoid this heavy style. Registered beans with a built-in lifecycle opened the door to wrapping these POJO-based objects with proxies that allowed the application of services.

One of Spring's earliest features was transactional support. Due to Spring's revolutionary nature of registering something such as `VideoService` with the application context, you could easily wrap a bean with a proxy that applied Spring's `TransactionTemplate` to every method call made from an outside caller.

This made it easy to unit test `VideoService`, ensuring it did its job while making transactional support a configuration step that the service didn't even have to know about.

When Java 5 emerged with its support for annotations, it became even easier to apply things. Transactional support could be applied with a simple `@Transaction` annotation.

By keeping services light and POJO-oriented, Spring effected a lightening of development.

Perhaps the application of annotations (and nothing else) is debatable as to whether it's really a POJO. But the idea of having to verify an application built up out of POJOs lets us build confidence in our system faster.

With all this prep work done, it's time to dig and start writing queries, as we'll see in the next section!

Creating a Spring Data repository

What is the best query? The one we don't have to write!

This may sound absurd, but Spring Data really makes it possible to write *lots* of queries…without writing them. The simplest one is based on the repository pattern.

This pattern was originally published in *Patterns of Enterprise Application Architecture, Martin Fowler, Addison-Wesley*.

A repository essentially gathers all the data operations for a given domain type in one place. The application talks to the repository in *domain speak*, and the repository in turn talks to the data store in *query speak*.

Before Spring Data, we had to write this translation of action by hand. But Spring Data provides the means to read the metadata of the data store and perform **query derivation**.

Let's check it out. Create a new interface called VideoRepository.java, and add the following code:

```
public interface VideoRepository extends JpaRepository
    <VideoEntity, Long> {
}
```

The preceding code can be explained as follows:

- It extends JpaRepository with two generic parameters: VideoEntity and Long (the domain type and the primary key type)
- JpaRepository, a Spring Data JPA interface, contains a set of already supported **Change Replace Update Delete** (**CRUD**) operations

Believe it or not, this is all we need to get going.

One of the most important things to understand is that by peeking inside of JpaRepository using our IDE, we'll discover that this class hierarchy ends with Repository. This is a marker interface with nothing inside it.

Spring Data is coded to look for all Repository instances and apply its various query derivation tactics. This means that any interface we create that extends Repository or any of its subinterfaces will be picked up by Spring Boot's component scanning and automatically registered for us to use.

But that is not all. JpaRepository comes loaded with various ways to fetch data, as shown with the following operations:

- findAll(), findAll(Example<S>), findAll(Example<S>, Sort), findAll(Sort), findAllById(Iterable<ID>), findById(ID), findAll(Pageable), findAll(Example<S>, Pageable),

```
findBy(Example<S>), findBy(Example<S>, Pageable),
findBy(Example<S>, Sort), findOne(Example<S>)
```

- `deleteById(ID)`, `deleteAll(Iterable<T>)`, `deleteAllById(Iterable<ID>)`, `deleteAllByIdInBatch(Iterable<ID>)`, `deleteAllInBatch()`

- `save(S)`, `saveAll(Iterable<S>)`, `saveAllAndFlush(Iterable<S>)`, `saveAndFlush(S)`

- `count()`, `count(Example<S>)`, `existsById(ID)`

These aren't all found directly in `JpaRepository`. Some are in other Spring Data repository interfaces further up the hierarchy including `ListPagingAndSortingRepository`, `ListCrudRepository`, and `QueryByExampleExecutor`.

The generic types sprinkled in the various signatures may seem a little confusing. Check out the following list to decode them:

- `ID`: The generic type of the repository's primary key
- `T`: The generic type of the repository's immediate domain type
- `S`: The generic subtype that extends `T` (sometimes used for **projection types**)

There are some container types that are also used in many places. These can be described as follows:

- `Iterable`: An iterable type that does not require all its elements to be fully realized in memory
- `Example`: An object used to serve Query By Example

As we work our way through this chapter, we'll cover these various operations and how we can use them to create a powerpack of data access.

While all these operations provide an incredible amount of power, one thing they lack is the ability to query with more specific criteria. The next section provides the means to start crafting more detailed queries.

Using custom finders

To create a custom finder, go back to the repository we created earlier, `VideoRepository`, and add the following method definition:

```
List<VideoEntity> findByName (String name);
```

The preceding code can be explained as follows:

- The `findByName(String name)` method is called a **custom finder**. We never have to implement this method. Spring Data will do it for us as described in this section.

- The return type is `List<VideoEntity>`, indicating it must return a list of the repository's domain type.

This interface method is all we need to write a query. The magic of Spring Data is that it will parse the method name. All repository methods that start with `findBy` are flagged as queries. After that, it looks for field names (`Name`) with some optional qualifiers (`Containing` and/or `IgnoreCase`). Since this is a field, it expects there to be a corresponding argument (`String name`). The name of the argument doesn't matter.

Spring Data JPA will literally translate this method signature into `select video.* from VideoEntity video where video.name = ?1`. As a bonus, it even performs proper binding on the incoming argument to avoid SQL injection attacks. It will convert every row coming back into a `VideoEntity` object.

> **Tip**
> What are SQL injection attacks? Anytime you give system users the chance to enter a piece of data and have it spliced into a query, you run the risk of someone inserting bits of SQL to maliciously attack the system. In general, blindly copying and pasting user inputs with production queries is a risky move. Binding arguments provide a much safer approach, forcing all user inputs to come in the data store's front door and get applied properly to query creation.

The ability to write type-safe queries based on domain types cannot be overstated. We also don't need to cope with the names of tables or their columns. Spring Data will use all the built-in metadata to craft the SQL needed to talk to our relational database.

On top of that, because this is JPA, we don't even have to sweat database dialects. Whether we are talking to MySQL, PostgreSQL, or some other instance, JPA will largely handle those idiosyncrasies.

There are additional operators used by custom finders:

- And and `Or` can be used to combine multiple property expressions. You can also use `Between`, `LessThan`, and `GreaterThan`

- You can apply `IsStartingWith`, `StartingWith`, `StartsWith`, `IsEndingWith`, `EndingWith`, `EndsWith`, `IsContaining`, `Containing`, `Like`, `IsNotContaining`, `NotContaining`, and `NotContains`

- You can apply `IgnoreCase` against a single field, or if you want to apply it to all properties, use `AllIgnoreCase` at the end

- You can apply `OrderBy` with `Asc` or `Desc` against a field when you know the ordering in advance

> **Note**
>
> `Containing` versus `StartsWith` versus `EndsWith` versus `Like`
>
> In **Jakarta Persistence Query Language (JPQL)** , `%` is a wildcard you can use for doing a partial match with `LIKE`. To apply it yourself, just apply `Like` onto the finder, for example, `findByNameLike()`. But if you're doing something simple, like putting the wildcard at the beginning, just use `StartsWith` and provide the partial token. Spring Data will plug in the wildcard for you. `EndsWith` puts the wildcard at the end and `Containing` puts one on each side. If you need something more complicated, then `Like` puts you in control, as in `%firstthis%thenthis%`.

Custom finders can also navigate relationships. For example, if the repository's domain type were `Person` and it had an `Address` field with `ZipCode`, we could write a custom finder called `findByAddressZipCode(ZipCode zipCode)`. This will generate a join to find the right results.

In the event that Spring Data runs into an ambiguous situation, it's possible to resolve things. For example, if that `Person` object just mentioned also had an `addressZip` field, Spring Data would naturally take that over navigating across a relationship. To force it to navigate properly, use an underscore (_) like this: `findByAddress_ZipCode(ZipCode zipCode)`.

Assuming we wanted to apply some of these techniques, what about creating a search box for our web app from the previous chapter?

Let's add a search box to the Mustache template we created in *Chapter 2, Creating a Web Application with Spring Boot*, `index.mustache`, as follows:

```
<form action="/multi-field-search" method="post">
  <label for="name">Name:</label>
  <input type="text" name="name">
  <label for="description">Description:</label>
  <input type="text" name="description">
  <button type="submit">Search</button>
</form>
```

The preceding code can be described as follows:

- The action denotes `/multi-field-search` as the target URL, with an HTTP method of `POST`
- There is a label and a text input for both `name` and `description`
- The button labeled `Search` will actuate the whole form

When the user enters search criteria in either box and clicks **Submit**, it will `POST` a form to `/multi-field-search`.

To handle this, we need a new method in our controller class that can parse this. As mentioned in the previous chapter, Mustache needs a data type to collect the name and description fields. A Java 17 record is perfect for defining such lightweight data types.

Create `VideoSearch.java` and add the following code:

```
record VideoSearch(String name, String description) {
}
```

This Java 17 record has two `String` fields—`name` and `description`—that perfectly match up with the names defined earlier in the HTML form.

Using this data type, we can add another method to `HomeController` from *Chapter 2, Creating a Web Application with Spring Boot*, to process the search request:

```
@PostMapping("/multi-field-search")
public String multiFieldSearch( //
  @ModelAttribute VideoSearch search, //
  Model model) {
  List<VideoEntity> searchResults = //
    videoService.search(search);
  model.addAttribute("videos", searchResults);
  return "index";
}
```

The preceding controller method can be described as follows:

- `@PostMapping("/multi-field-search")` is Spring MVC's annotation to mark the method for processing HTTP POST requests to the URL.

- The search argument has the `VideoSearch` record type. The `@ModelAttribute` annotation is Spring MVC's signal to deserialize the incoming form. The `Model` argument is a mechanism to send information out for rendering.

- There is a newly minted `search()` method where our `VideoSearch` criteria get forwarded to `VideoService` (which we'll define further down). The results are inserted into the `Model` object under the name `videos`.

- The method finally returns the name of the template to render, `index`. As a refresher from the previous chapter, Spring Boot is responsible for translating this name to `src/main/resources/templates/index.mustache`.

In defining the web method for handling search requests, we must now design a `VideoService` method to do a search. This is where it gets a little tricky. So far, we've simply ferried the details of the request.

Now, it's time to make the request, and all kinds of things could happen:

- The user could have entered both **name** *and* **description** details
- The user could have entered only the **name** field
- The user could have entered only the **description** field

What information the user enters could swing things one way or the other. For example, if the name field were empty, we don't want to attempt to match against an empty string, because that would match everything.

We need a method signature as follows:

```
public List<VideoEntity> search(VideoSearch videoSearch)
```

The preceding code meets our obligation of taking a `VideoSearch` input and translating it to a list of `VideoEntity` objects.

From here, we need to switch to using a both name and description, based on the inputs, with the first path being as follows:

```
if (StringUtils.hasText(videoSearch.name())
  && StringUtils.hasText(videoSearch.description())) {
  return repository
    .findByNameContainsOrDescriptionContainsAllIgnoreCase(
      videoSearch.name(), videoSearch.description());
}
```

The preceding code has some key components:

- `StringUtils` is a Spring Framework utility that lets us check that both fields of the `VideoSearch` record actually have some text and are neither empty nor `null`.
- Assuming both fields are populated, we can then invoke a custom finder that matches the name field and the `description` field, but with the `Contains` qualifiers and the `AllIgnoreCase` modifier. Basically, we're looking for a partial match on both fields, and the casing shouldn't be an issue.

If either field is empty (or `null`), then we need additional checks, as follows:

```
if (StringUtils.hasText(videoSearch.name())) {
  return repository.findByNameContainsIgnoreCase
    (videoSearch.name());
}


if (StringUtils.hasText(videoSearch.description())) {
  return repository.findByDescriptionContainsIgnoreCase
    (videoSearch.description());
}
```

The preceding code is somewhat similar, but varies:

- Using the same `StringUtils` utility method as before, check if the name field has text. If so, invoke the custom finder matching on `name` with the `Contains` and `IgnoreCase` qualifiers.

- Also, check whether the `description` field has text. If so, use the custom finder that matches on `description` with `Contains` and `IgnoreCase` qualifiers.

Finally, if both fields are empty (or `null`), there is but one result to return:

```
return Collections.emptyList();
```

Since this is the final state of what can happen, there is no need for an `if` clause. If our code has made it here, there is nothing to do but return an empty list.

If you felt that this series of `if` clauses is a little clunky, I'd agree! There are yet more ways to query the database using Spring Data JPA, and we'll investigate further. Later in this chapter, we'll see how we can use some of these tactics to our advantage and see about making a slicker, smoother solution.

Sorting the results

There are several ways to sort data. We just mentioned adding an `OrderBy` clause earlier in this chapter. This is a static approach, but it's also possible to delegate this to the caller.

Any custom finder can also have a `Sort` parameter, allowing the caller to decide how to sort the results:

```
Sort sort = Sort.by("name").ascending()
  .and(Sort.by("description").descending());
```

This fluent `Sort` API lets us build up a series of columns and allows us to choose whether these columns should be sorted in ascending or descending order. This is also the order in which the sorting will be applied.

If you're worried about using string values to represent columns, then ever since the days of Java 8, Spring Data has also supported strongly-typed sort criteria, as follows:

```
TypedSort<Video> video = Sort.sort(Video.class);
Sort sort = video.by(Video::getName).ascending()
   .and(video.by(Video::getDescription).descending());
```

Limiting query results

There are several ways to constrain the results. Why is this needed? Just imagine querying a table with 100,000 rows. Don't want to fetch *all* of that, right?

Some of the options we can apply to custom finders include the following:

- `First` or `Top`: Finds the first entry in the result set, for example, `findFirstByName(String name)` or `findTopByDescription(String desc)`.
- `FirstNNN` or `TopNNN`: Finds the first *NNN* entries in the result set, for example, `findFirst5ByName(String name)` or `findTop3ByDescription(String desc)`.
- `Distinct`: Apply this operator for data stores that support it, for example, `findDistinctByName(String name)`.
- `Pageable`: Request a page of data, for example, `PageRequest.of(1, 20)` will find the first page (0 being the first page) with a page size of 20. It's also possible to provide a `Sort` parameter to `Pageable`.

It's also important to point out that not only can we write custom finders, but we can also write a custom `existsBy`, `deleteBy`, and `countBy` method. They all support the same conditions described in this section.

Look at the following set of examples:

- `countByName(String name)`: Runs the query but with a `COUNT` operator applied, returning an integer instead of the domain type
- `existsByDescription(String description)`: Runs the query but gleans whether or not the result is empty
- `deleteByTag(String tag)`: `DELETE` instead of `SELECT`

> **Tip**
>
> *SQL vs. JPQL* – What query does Spring Data JPA actually write? JPA provides a construct to build up queries known as `EntityManager`. `EntityManager` provides APIs to assemble queries using JPQL. Spring Data JPA parses methods from the repository method and talks to `EntityManager` on your behalf. Under the covers, JPA is responsible for converting JPQL to **Structured Query Language (SQL)**, the language relational data stores speak. Certain concepts, like injection attacks, don't really matter whether we are talking about JPQL or SQL. But when it comes down to the actual queries, it's important to be sure we're talking about the right thing.

Custom finders are incredibly powerful. They make it possible to capture business concepts rapidly without fiddling with writing queries.

But there's a fundamental trade-off that may not make them ideal for every situation.

Custom finders are almost completely fixed. True, we can provide custom criteria through the arguments, and it's possible to dynamically adjust sorting and paging. But the columns we choose for the criteria and how they are combined (`IgnoreCase`, `Distinct`, and so on) are fixed when we write them.

We've seen a limitation of this in the previous search box scenario. Simply having two parameters, `name` and `description`, sent us down a path to write a series of `if` clauses to pick the right custom finder.

Imagine how this would explode if we added more and more options. To cut to the chase, this approach results in a combinatorial explosion of finder methods to cover it all and the `if` statements quickly become long-winded and a little hard to reason about. What happens if we add another field?

The problem, as stated, is that custom finders are rather static in the criteria we can apply. Thankfully, Spring Data offers a way out of such fluid situations, which we'll cover in the next section.

Using Query By Example to find tricky answers

So, what happens when the exact criteria for a query vary from request to request? In short, we need a way to feed Spring Data an object that captures the fields we're interested in while ignoring the ones that we aren't.

The answer is **Query By Example**.

Query By Example lets us create a **probe**, which is an instance of the domain object. We populate the fields with criteria we want to apply and leave the ones we aren't interested in empty (`null`).

We then wrap the probe, creating an `Example`. Check out the following example:

```
VideoEntity probe = new VideoEntity();
probe.setName(partialName);
probe.setDescription(partialDescription);
```

```
probe.setTags(partialTags);
Example<VideoEntity> example = Example.of(probe);
```

The preceding code can be broken down as follows:

- The first few lines are where we create the probe, presumably pulling down fields from a Spring MVC web method where they were posted, some populated, some `null`

- The last line wraps the `Example<VideoEntity>` probe with a policy of exactly matching only the non-`null` fields

Earlier, when discussing the calamity of custom finders (under the *Using customer finders* section), we mentioned applying an `AllIgnoreCase` clause. To do the same for Query By Example, we'd have to alter our example as follows:

```
Example<VideoEntity> example = Example.of(probe,
    ExampleMatcher.matchingAll()
      .withIgnoreCase()
      .withStringMatcher(StringMatcher.CONTAINING));
```

Assuming we were using the exact same probe as before, `ExampleMatcher` alters things as follows:

- It matches on all fields, essentially an And operation as before. However, if we wanted to switch to an Or operation, we could switch to `matchingAny()`.

- `withIgnoreCase()` tells Spring Data to make the query case insensitive. It essentially applies a `lower()` operation on all the columns (so adjust any indexes suitably!).

- `withStringMatcher()` applies a CONTAINING filter to make it a partial match on all non-`null` columns. Under the hood, Spring Data wraps each column with a wildcard and then applies the LIKE operator.

Assuming we put together `Example<VideoEntity>`, how do we use it? The `JpaRepository` interface we are leveraging comes with `findOne(Example<S> example)` and `findAll(Example<S> example)`.

> Tip
>
> `JpaRepository` inherits these `Example`-based operations from `QueryByExample Executor`. If you are rolling your own extension of `Repository`, you can either extend `QueryByExampleExecutor` or add the `findAll(Example<S>)` methods by hand. Either way, as long as the method signature is there, Spring Data will happily execute your Query By Example.

So far, we have looked at how some sort of search box or filter on the web page could be used to assemble a probe. If we decided to switch from a multi-field setup and instead had a universal search box where there is only one input, it would take little effort to adapt it!

Let's see if we can noodle out such a search box:

```
<form action="/universal-search" method="post">
  <label for="value">Search:</label>
  <input type="text" name="value">
  <button type="submit">Search</button>
</form>
```

This form in the preceding code is quite similar to the HTML template created earlier in this chapter, with the following exceptions:

- The target URL is /universal-search
- There is only one input, value

Again, to transport this piece of input data, we need to wrap it with a DTO. Thanks to Java 17 records, this is super simple. Just create a UniversalSearch record as follows:

```
record UniversalSearch(String value) {
}
```

The preceding DTO contains one entry: value.

To process this new UniversalSearch, we need a new web method:

```
@PostMapping("/universal-search")
public String universalSearch(
  @ModelAttribute UniversalSearch search, Model model) {
    List<VideoEntity> searchResults =
      videoService.search(search);
  model.addAttribute("videos", searchResults);
  return "index";
}
```

The preceding search handler is quite similar to the multi-field one we made earlier, with the following exceptions:

- It responds to /universal-search
- The incoming form is captured in the single-value UniversalSearch type

- The search DTO is passed on to a different `search()` method, which we'll write further down in this section

- The search results are stored in the `Model` field to be rendered by the `index` template

Now, we're poised to leverage Query By Example by creating a `VideoService.search()` method that takes in one value and applies it to all the fields, as follows:

```
public List<VideoEntity> search(UniversalSearch search) {
  VideoEntity probe = new VideoEntity();
  if (StringUtils.hasText(search.value())) {
    probe.setName(search.value());
    probe.setDescription(search.value());
  }
  Example<VideoEntity> example = Example.of(probe, //
    ExampleMatcher.matchingAny() //
      .withIgnoreCase() //
      .withStringMatcher(StringMatcher.CONTAINING));
  return repository.findAll(example);
}
```

The preceding alternative search method can be explained as follows:

- It takes in the `UniversalSearch` DTO.

- We create a probe based on the same domain type as the repository and copy the `value` attribute into the probe's `Name` and `Description` fields, but only if there is text. If the `value` attribute is empty, the fields are left `null`.

- We assemble an `Example<VideoEntity>` using the `Example.of` static method. However, in addition to providing the probe, we also provide additional criteria of ignoring the casing and applying a `CONTAINING` match, which puts wildcards on both sides of every input.

- Since we're putting the same criteria in all fields, we need to switch to `matchingAny()`, that is, an `Or` operation.

With a single design change to the UI and by switching to Query By Example, we were able to adjust the backend to find results.

This isn't just effective and maintainable, it's pretty easy to read and understand what's happening. If we start adding more attributes to this video-based structure, it doesn't look hard to adjust.

> **Tip**
>
> In case you're thinking you can simply create a finder that matches on all fields, providing `null` to the columns you want to ignore, this won't work. This is your friendly reminder that in relational databases, `null` doesn't equal `null`. That's why Spring Data also has `IsNull` and `IsNotNull` as qualifiers; for example, `findByNameIsNull` will find any entries where the name field is `null`.

However, that is not all. There are other ways to fashion queries, including a more fluent way, as shown in the next section.

Using custom JPA

If all else fails and we can't seem to bend Spring Data's query derivation tactics to meet our needs, it's possible to write the JPQL ourselves.

In our repository interface, we can create a query method as follows:

```
@Query("select v from VideoEntity v where v.name = ?1")
List<VideoEntity> findCustomerReport(String name);
```

The preceding method can be explained as follows:

- `@Query` is Spring Data JPA's way to supply a custom JPQL statement.
- It's possible to include positional binding parameters using `?1` to tie it to the `name` argument.
- Since we are providing the JPQL, the name of the method no longer matters. This is our opportunity to pick a better name than what custom finders constrained us to.
- Because the return type is `List<VideoEntity>`, Spring Data will form a collection.

Using `@Query` essentially sidesteps any query writing done by Spring Data and uses the user's supplied query with one exception: Spring Data JPA will still apply `ordering` and `paging`. Because SORT clauses can be appended at the end of queries, Spring Data JPA will let us provide a `Sort` argument and apply it.

While we are focusing on Spring Data JPA details such as JPQL, almost every other Spring Data module has a corresponding `@Query` annotation. Each data store lets us write custom queries in the data store's query language, for example: **MongoQL**, **Cassandra Query Language** (**CQL**), or even **Nickel/Couchbase Query Language** (**N1QL**).

In terms of Spring Data JPA, it must be stressed that this annotation lets us provide JPQL. Now, the previous example is a bit simplistic for a custom query. If you were thinking that based on what you've read up to this point, it would be a perfect candidate for `findByName(String name)`, you're right!

However, sometimes we have that custom query that requires joining *lots* of different tables. Maybe something more similar to the following:

```
@Query("select v FROM VideoEntity v " //
    + "JOIN v.metrics m " //
    + "JOIN m.activity a " //
    + "JOIN v.engagement e " //
    + "WHERE a.views < :minimumViews " //
    + "OR e.likes < :minimumLikes")
List<VideoEntity> findVideosThatArentPopular( //
    @Param("minimumViews") Long minimumViews, //
    @Param("minimumLikes") Long minimumLikes);
```

The preceding code can be explained as follows:

- This `@Query` shows a JPQL statement that joins four different tables together, using standard inner joins.

- `:minimumViews` and `:minimumLikes` are named parameters (instead of the default positional parameters). They are bound to the method arguments by the Spring Data `@Param("minimumViews")` and `@Param("minimumLikes")` annotations.

The preceding method is getting closer to something `@Query` is good for. A comparable custom finder would be `findByMetricsActivityViewsLessThanOrEngagementLikesLessThan(Long minimumViews, Long minimumLikes)`.

> **Tip**
> Choosing between custom finders and `@Query` is hard. To be honest, for this example where we join four tables together, I'd still take that custom finder, because I know it will be right. But as that finder method gets longer and longer, things begin to shift more favorably toward writing the query by hand. A key factor is the number of `WHERE` clauses as well as the number of complex (that is, outer) `JOIN` clauses. Essentially, the harder it gets to capture it in a simple name, the better it becomes to take control of the whole query.

And if JPQL is getting in the way, it's possible to even move past that and write pure SQL using `@Query`'s `nativeQuery=true` argument.

Spring Data JPA 3.0 includes **JSqlParser**, a **SQL-parsing library**, making it possible to write queries as follows:

```
@Query(value="select * from VIDEO_ENTITY where NAME = ?1",
nativeQuery=true)
List<VideoEntity> findCustomWithPureSql(String name);
```

Why do we write queries as shown in the preceding code? There are several possible reasons, actually:

- Perhaps we need access to a customer report, but all the relevant tables don't really connect to the stuff the rest of our finders operate on. Is it worth it to climb through configuring a stack of entity types for one report? It may be easier to focus on writing the pure SQL for the report.

- I've seen reports that literally join 20 tables with complex left outer joins, correlated subqueries, and other complexities. Converting it to JPQL for the sake of using JPA didn't make sense.

The criteria for swapping our custom finders for native SQL are pretty close to whether or not we would swap it out for custom JPQL. It really hinges on how comfortable we are with JPQL versus SQL.

> **Tip**
> Personally, if I were using @Query, I'd probably switch to pure SQL, not JPQL. But that's probably because I have many more years of experience with SQL than with JPQL. Having worked on a 24x7 system with five 9s of availability with over 200 queries, I can write left outer joins and correlated subqueries in my sleep. The JPQL equivalent would require too much study. But, perhaps JPQL is your thing. If you jam in the land of JPQL, then run with that. Whatever gets the job done, go for it!

Another factor to consider is that Spring Data JPA doesn't support dynamic sorting when doing native queries. Doing so would require manipulating the SQL by adding the SORT clauses. It is possible to support paging requests with a `Pageable` argument. But it requires that we also fill in @Query's `countQuery` entry, providing the SQL statement to count. (Spring Data JPA can iterate over the result set, providing pages of results.)

It's also important to understand that Spring Data will still handle connection management and transaction handling.

Summary

Over the course of this chapter, we have learned a multitude of ways to fetch data using Spring Data JPA. We then hooked several variations of queries into some search boxes. We used Java 17 records to quickly assemble DTOs to ferry form requests into web methods and onto `VideoService`.

We tried to assess when it makes sense to use various querying tactics.

In the next chapter, *Securing an Application with Spring Boot*, we'll explore how to lock down our application and get it ready for production.

4

Securing an Application with Spring Boot

In the previous chapter, we learned how to query for data using Spring Data JPA. We figured out how to write custom finders, use Query By Example, and even how to directly access the data store custom JPQL and SQL.

In this chapter, we'll see how to keep our application secure.

Security is a critical issue. I have said, multiple times, that your application isn't real until it's secured.

But security isn't just a switch we flip and we're done. It's a complex problem that requires multiple layers. It requires careful respect.

If there is one thing to appreciate as we dive into this chapter, it's to never attempt to secure things on your own. Don't roll your own solution. Don't assume it's easy. The person who wrote the commercial utility to crack Word documents for users who had lost their password said he introduced a deliberate delay so it didn't appear instantaneous.

There are security engineers, computer scientists, and industry leaders who have studied application security for years. Adopting tools and practices developed by industry experts is the first step toward ensuring our application's data and its users are properly protected.

That's why our first step will be to turn to a well-respected security tool: **Spring Security**.

Spring Security has been developed in the open from the beginning (2003). This framework isn't proprietary but instead has had contributions from well-respected security professionals around the globe. Also, it's actively maintained by a dedicated portion of the Spring team.

In this chapter, we'll cover the following topics:

- Adding Spring Security to a Spring Boot application
- Creating our own users with a custom security policy

- Swapping hardcoded users with a Spring Data-backed set of users

- Leveraging Google to authenticate users

- Securing web routes and HTTP verbs

- Securing Spring Data methods

> **Where to find this chapter's code**
>
> The source code for this chapter can be found at `https://github.com/PacktPublishing/Learning-Spring-Boot-3.0/tree/main/ch4`.

Adding Spring Security to our project

Before we can do anything with Spring Security, we must add it to our project.

To add Spring Security to our already drafted app, we can easily use the same tactic from the previous chapters:

1. Visit `start.spring.io`.

2. Enter the same project artifact details as before.

3. Click on **DEPENDENCIES**.

4. Select **Spring Security**.

5. Click on **EXPLORE**.

6. Look for the `pom.xml` file and click on it.

7. Copy the new bits onto the clipboard. Watch out! Spring Security has both a starter as well as a test module.

8. Open up our previous project inside our **IDE**.

9. Open our `pom.xml` file and paste the new bits into the right places.

Hit the refresh button in the IDE, and we're ready to go!

Out of the box, we can run the application we have built up to this point. The exact same web app backed by Spring Data JPA is runnable...and it will be locked down.

Kind of.

When Spring Boot detects Spring Security on the path, it locks everything down with a randomly generated password. This can be a good thing...or a bad thing.

If we were giving a quick rundown to the CTO of a company, showing how the power of Spring Boot allows us to lock down applications, this can be a good thing.

But if we intend to do anything deeper than pitching, we need another approach. The problem with using Spring Boot's autoconfigured "user" username and random password is that the password changes every time the app restarts.

We could override the username, password, and even the roles using `application.properties`, but this isn't scalable. There is another approach that takes about the same amount of effort and sets us up for a more realistic approach.

This is what we'll tackle next.

Creating our own users with a custom security policy

Spring Security has a highly pluggable architecture, which we shall take full advantage of throughout this chapter.

The key aspects of securing any application are as follows:

- Defining the source of users
- Creating access rules for the users
- Associating various parts of the app with the access rules
- Applying the policy to all aspects of the application

Let's start with the first step and create a source of users. Spring Security comes with an interface for just this task: `UserDetailsService`.

To leverage this, we'll start by creating a `SecurityConfig` Java class with the following code:

```
@Configuration
public class SecurityConfig {

  @Bean
  public UserDetailsService userDetailsService() {
    UserDetailsManager userDetailsManager =
      new InMemoryUserDetailsManager();
    userDetailsManager.createUser(
      User.withDefaultPasswordEncoder()
        .username("user")
        .password("password")
        .roles("USER")
        .build());
    userDetailsManager.createUser(
```

```
        User.withDefaultPasswordEncoder()
            .username("admin")
            .password("password")
            .roles("ADMIN")
            .build());
    return userDetailsManager;
  }
}
```

The preceding security code can be described as follows:

- @Configuration is Spring's annotation to signal that this class is a source of bean definitions rather than actual application code. Spring Boot will detect it through its component scanning and automatically add all its bean definitions to the application context.

- UserDetailsService is Spring Security's interface for defining a source of users. This bean definition, marked with @Bean, creates InMemoryUserDetailsManager.

- Using InMemoryUserDetailsManager, we can then create a couple of users. Each user has a *username*, a *password*, and a *role*. This code fragment also uses the withDefaultPasswordEncoder() method to avoid encoding the password.

What's important to understand is that when Spring Security gets added to the classpath, Spring Boot's autoconfiguration will activate Spring Security's @EnableWebSecurity annotation. This switches on a standard configuration of various filters and other components.

The components are dynamically picked based on whether this is **Spring MVC** or its reactive variant, **Spring WebFlux**. One of the beans required is a UserDetailsService bean.

Spring Boot will autoconfigure the single-user instance version we talked about in the preceding section. But, because we defined our own, Spring Boot will back off and let our version take its place.

> **Tip**
>
> In the code so far, we used withDefaultPasswordEncoder() to store passwords in the clear. Do NOT do this in production! Passwords need to be encrypted before being stored. In fact, there is a long and detailed history of the proper storage of passwords that reduces the risk of not just sniffing out a password but guarding against dictionary attacks. See https://springbootlearning.com/password-storage for more details on properly securing passwords when using Spring Security.

Believe it or not, this is enough to run the application we've been developing in this book! Go ahead, either right-click on the **public static void main()** method and run it or use ./mvnw spring-boot:run in the terminal.

Once the app is up, visit `localhost:8080` in a new browser tab, and you should be automatically redirected to this page at `/login`:

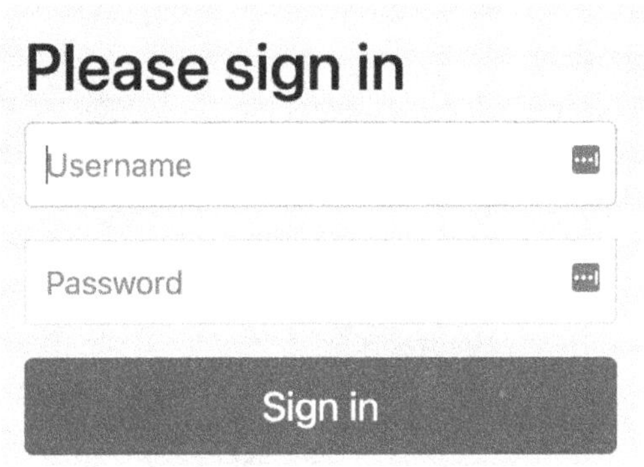

Figure 4.1 – Spring Security's default login form

This is Spring Security's built-in login form. No need to roll our own. If we enter one of the accounts from the `userDetailsService` bean, it will let us in!

If hard-coded passwords are making you a tad nervous, then check out the next section, where we'll move the storage of user credentials to an external database.

Swapping hardcoded users with a Spring Data-backed set of users

Creating a hardcoded set of users is great if we're creating a demo (or writing a book!), but it's no way to build a real, production-oriented application. Instead, it's better to outsource user management to an external database.

By having the application reach out and authenticate against an external user source, it makes it possible for another team, such as our security engineering team, to manage the users through a completely different tool that manages that database.

Decoupling user management from user authentication is a great way to improve the security of the system. So, we'll combine some of the techniques we learned in the previous chapter with the `UserDetailsService` interface we learned about in the previous section.

Since we already have Spring Data JPA and H2 on the classpath, we can start off by defining a JPA-based `UserAccount` domain object as follows:

```
@Entity
public class UserAccount {
    @Id
```

```
    @GeneratedValue
    private Long id;
    private String username;
    private String password;
    @ElementCollection(fetch = FetchType.EAGER)
    private List<GrantedAuthority> authorities = //
      new ArrayList<>();
}
```

The preceding code contains some key features:

- As discussed in *Chapter 3*, *Querying for Data with Spring Boot*, @Entity is JPA's annotation for denoting classes that are mapped onto relational tables.

- The primary key is marked by the @Id annotation. @GeneratedValue signals the JPA provider to generate unique values for us.

- This class also has a username, a password, and a list of authorities.

- Because the authorities are a collection, JPA 2 offers a simple way to handle this using their @ ElementCollection annotation. All these authority values will be stored in a separate table.

Before we can ask Spring Security to fetch user data, we should probably load some up. While in production, we would need to build up a separate tool to create and update the tables. For now, we can just pre-load some entries directly.

To do this, let's create a Spring Data JPA repository definition aimed at the user manager by creating a UserManagementRepository interface as follows:

```
public interface UserManagementRepository extends
  JpaRepository<UserAccount, Long> {
}
```

The preceding repository extends Spring Data JPA's JpaRepository, providing a whole suite of operations needed by any user-managing tool.

To take advantage of it, we need Spring Boot to run a piece of code when the app starts up. Add the following bean definition to our SecurityConfig class:

```
@Bean
CommandLineRunner initUsers(UserManagementRepository
  repository) {
  return args -> {
    repository.save(new UserAccount("user", "password",
```

```
      "ROLE_USER"));
    repository.save(new UserAccount("admin", "password",
      "ROLE_ADMIN"));
  };
}
```

The preceding bean defines Spring Boot's `CommandLineRunner` (through a *Java 8* **lambda function**).

> **Tip**
> `CommandLineRunner` is a **single abstract method** (**SAM**), meaning it has just one method that must be defined. This trait lets us instantiate `CommandLineRunner` using a **lambda expression** instead of creating an anonymous class from the pre-Java 8 days.

In our example, we have a bean definition that depends upon `UserManagementRepository`. Inside the lambda expression, this repository is used to save two `UserAccount` entries: one user and one admin. With these entries in place, we can finally code our JPA-oriented `UserDetailsService`!

To fetch a `UserAccount` entry, we need another Spring Data repository definition. Only this time, we need a very simple one. Nothing that involves saving or deleting. So, create an interface called `UserRepository` as follows:

```
public interface UserRepository extends
  Repository<UserAccount, Long> {
      UserAccount findByUsername(String username);
}
```

The preceding code is different from the previous repository we created (`UserManagementRepository`) in the following ways:

- It extends `Repository` instead of `JpaRepository`. This means it starts with absolutely nothing. There are no operations defined except what is right here.
- It has a custom finder, `findByUsername`, used to fetch a `UserAccount` entry based on the username. This is exactly what we'll need to serve Spring Security later in this section.

This is one of the places where Spring Data really shines. We have focused on the domain of `UserAccount` and written one repository that is involved with storing data while defining another repository focused on fetching a single entry. All without getting dragged down into writing JPQL or SQL.

With all this in place, we can finally create a bean definition that lets us replace the `UserDetailsService` bean we defined in the previous section by adding this to our `SecurityConfig` class:

```
@Bean
UserDetailsService userService(UserRepository repo) {
  return username -> repo.findByUsername(username)
    .asUser();
}
```

The preceding bean definition calls for `UserRepository`. We then use it to fashion a lambda expression that forms `UserDetailsService`. If we peek at that interface, we'll find it's another SAM, a single method named `loadUserByName`, transforming a string-based `username` field into a `UserDetails` object. The incoming argument is `username`, which we can then delegate to the repository.

`UserDetails` is Spring Security's representation of a user's information. This includes `username`, `password`, `authorities`, and some Boolean values representing locked, expired, and enabled.

Let's slow down and double-check things, because that last bit may have flown by a bit quickly. The `userService` bean in the preceding code produces a `UserDetailsService` bean, not the `UserDetails` object itself. It's a service meant to retrieve user data.

The lambda expression inside the bean definition gets transformed into `UserDetailsService.loadUserName()`, a function that takes `username` for input and produces a `UserDetails` object as its output. Imagine someone entering their username at a login prompt. This value is what gets fed into this function.

The repository does the key step of fetching a `UserAccount` from the database based on username. For Spring Security to work with this entity from the database, it must be converted to a Spring Security `User` object (which implements the `UserDetails` interface).

So, we need to circle back to `UserAccount` and add a convenience method, `asUser()`, to convert it:

```
public UserDetails asUser() {
  return User.withDefaultPasswordEncoder() //
    .username(getUsername()) //
    .password(getPassword()) //
    .authorities(getAuthorities()) //
    .build();
}
```

This method simply creates a Spring Security `UserDetails` object, using its builder and plugging in the attributes from our entity type. Now, we have a complete solution that outsources user management to a database table.

> **Warning**
>
> If you're worried about encoding passwords to prevent hack attacks, your concern is justified! This needs to be handled by the user management tool that actually stores these passwords, which we alluded to earlier. We also need to cope with the need to update roles. On top of that, we need a secure solution that guards against hash table attacks.

In fact, user management can become a tedious task to manage. Later in this chapter, in the *Leveraging Google to authenticate users* section, we'll investigate alternative ways to manage users.

We mentioned at the beginning of this section that we needed to define a source of users. Mission accomplished! The next bullet point to tackle is to define some access roles. So, let's delve into that in the next section.

Securing web routes and HTTP verbs

Locking down an application and only allowing authorized users to access it is a big step forward. But, it's seldom enough.

We must actually confine who can do what. So far, the process we've applied where people must prove their identity as part of a closed list of users is known as **authentication**.

But, the next piece of security that must be applied to any real application is what's called **authorization**, that is, what a user is allowed to do.

Spring Security makes this super simple to apply. The first step in customizing our security policy is to add one more bean definition to our `SecurityConfig` class created earlier in this chapter under the *Creating our own users with a custom security policy* section.

Up to this point, Spring Boot has had an autoconfigured policy in place. In fact, it may be simpler to show what Spring Boot has inside its own `SpringBootWebSecurityConfiguration`:

```
@Bean
SecurityFilterChain defaultSecurityFilterChain
  (HttpSecurity http) throws Exception {
    http.authorizeRequests().anyRequest().authenticated();
    http.formLogin();
    http.httpBasic();
```

```
    return http.build();
}
```

The preceding code fragment can be described as follows:

- @Bean signals this method as a bean definition to be picked up and added to the application context.

- SecurityFilterChain is the bean type needed to define a Spring Security policy.

- To define such a policy, we ask for a Spring Security HttpSecurity bean. This gives us a handle on defining rules that will govern our application.

- authorizeRequests defines exactly how we will, you know, *authorize requests*. In this case, any request is allowed if the user is authenticated and that is the only rule applied.

- In addition to that, the formLogin and httpBasic directives are switched on, enabling both **HTTP Form** and **HTTP Basic**, two standard authentication mechanisms.

- The HttpSecurity builder, with these settings, is then used to render SecurityFilterChain through which all servlet requests will be routed.

To provide more detail about formLogin and httpBasic, it's important to understand some facts.

Form authentication involves having a nice HTML form, which can be stylized to match the theme of the web application. Spring Security even provides a default one (which this chapter shall use). Form authentication also supports logging out.

Basic authentication has nothing to do with HTML and form rendering but instead involves a popup baked into every browser. There is no support for customization, and the only way to throw away credentials is to either shut down or restart the browser. Basic authentication also makes it simple to authenticate with a command-line tool, such as **curl**.

In general, by having both form and basic authentication, the application will defer to form-based authentication in the browser while still allowing basic authentication from command-line tools.

This security policy provided by Spring Boot contains no authorization whatsoever. It essentially grants access to everything as long as the user is authenticated.

An example of a more detailed policy could be as follows:

```
@Bean
SecurityFilterChain configureSecurity(HttpSecurity http)
  throws Exception {
    http.authorizeHttpRequests() //
        .requestMatchers("/resources/**", "/about", "/login")
          .permitAll() //
          .requestMatchers(HttpMethod.GET, "/admin/**")
```

```
        .hasRole("ADMIN") //
        .requestMatchers("/db/**").access((authentication,
          object) -> {
            boolean anyMissing = Stream.of("ADMIN",
                                            "DBA")//
              .map(role -> hasRole(role)
              .check(authentication, object).isGranted()) //
              .filter(granted -> !granted) //
              .findAny() //
              .orElse(false); //
      return new AuthorizationDecision(!anyMissing);
        }) //
        .anyRequest().denyAll() //
        .and() //
        .formLogin() //
        .and() //
        .httpBasic();
      return http.build();
}
```

The preceding security policy has a lot more detail, so let's take it apart, clause by clause:

- The method signature is identical to Spring Boot's default policy shown previously. The method name may be different, but that doesn't really matter.

- This policy uses `authorizeHttpRequests`, signaling web-based checks.

- The first rule is a path-based check to see if the URL starts with `/resources`, `/about`, or `/login`. If so, access is immediately granted, regardless of authentication status. In other words, these pages are freely accessible without logging in.

- The second rule looks for any `GET` calls to the `/admin` pages. These calls indicate that the user has the `ADMIN` role. This is where an HTTP verb can be combined with a path to control access. This can be especially useful to lock down things such as the `DELETE` operations.

- The third rule shows a much more powerful and customizable check. If the user is attempting to access anything underneath the `/db` path, then a special access check is performed. The preceding code has a lambda function where we are handed a copy of the current user's authentication along with the *object* that is being checked. The function takes a stream of roles (`DBA` and `ADMIN`), checks if the user is granted the role, looks for any roles not granted, and if there are any, denies access. In other words, the user must be both a `DBA` and `ADMIN` to access this path.

- The last rule denies access. This is a generally good pattern. If the user can't meet any of the earlier rules, they shouldn't be granted access to anything.

- After the rules, both form authentication and basic authentication are enabled, just like with Spring Boot's default policy.

Security is a complex beast. That's why no matter what rules are provided by Spring Security, we must always have the ability to write custom access checks. The rule governing access to /db/** is a perfect example.

Instead of expecting Spring Security to capture every permutation of possible rules, it's easier to grant us the ability to write a custom check. In our example, we have chosen to check that someone possesses all roles. (It should be noted that Spring Security has built-in functions to detect whether a user has *any one of* a given list of roles but cannot check for *all roles*.)

This custom rule we've written up is a perfect example of why many, many test cases should be written! We'll dig into this in more detail in *Chapter 5, Testing with Spring Boot*, but the complexity of these rules should serve as a prelude to why it's critical to test both success and failure paths, ensuring the rules are working!

> **Note**
>
> **Roles** versus **authorities**: Spring Security has a fundamental concept known as authorities. Essentially, an *authority* is a defined permission to access something. However, the concept of having ROLE_ADMIN, ROLE_USER, ROLE_DBA, and others where the prefix of ROLE_ categorizes such authorities is so commonly used that Spring Security has a full suite of APIs to support **role checking**. In this situation, a user who has the authority of ROLE_ADMIN or simply the role of ADMIN would be able to GET any of the admin pages.

Using what we've learned, let's see if we can fashion a suitable policy for our video-listing site. First, let's write down some requirements:

- Everyone must log in to access anything

- The initial list of videos should only be visible to authenticated users

- Any search features should be available to authenticated users

- Only admin users can add new videos

- Any other forms of access will be disabled

- These rules should apply to both the HTML web page as well as to command-line interactions

Using these requirements, we should be able to define a `SecurityChainFilter` bean:

```
@Bean
SecurityFilterChain configureSecurity(HttpSecurity http)
  throws Exception {
    http.authorizeHttpRequests() //
      .requestMatchers("/login").permitAll() //
      .requestMatchers("/", "/search").authenticated() //
      .requestMatchers(HttpMethod.GET, "/api/**")
      .authenticated() //
      .requestMatchers(HttpMethod.POST, "/new-video",
                       "/api/**").hasRole("ADMIN") //
      .anyRequest().denyAll() //
      .and() //
      .formLogin() //
      .and() //
      .httpBasic();
    return http.build();
}
```

The preceding security policy can be described as follows:

- `@Bean` denotes this bean definition as taking an `HttpSecurity` bean and producing a `SecurityChainFilter` bean. This is the hallmark of defining a Spring Security policy.

- Using `authorizeHttpRequests`, we can see a series of rules. The first one grants everyone access to the `/login` page, whether or not they are logged in.

- The second rule grants access to the base URL `/` as well as the search results to anyone who is authenticated. While we could have a specific role, there is no need to constrain the base page as such.

- The third rule confines `GET` access to any `/api` URL to an authenticated user. This allows command-line access to the site and is the API equivalent of allowing any authenticated user access to the base web page.

- The fourth rule restricts `POST` access to both `/new-video` as well as `/api/new-video` only to authenticated users who also have the `ADMIN` role.

- The fifth rule says that any user that doesn't match any of the previous rules will be denied access regardless of authentication or authorization.

- To wrap things up, both form and basic authentication are enabled.

There is one lingering issue we must decide on: **Cross-Site Request Forgery** (**CSRF**). We'll make our choice on what path to take in the next section.

To CSRF or not to CSRF, that is the question

CSRF represents a particular attack vector that Spring Security guards against, by default. (Spring Security guards against many attack methods, but most don't require a policy decision).

It's a bit technical, but CSRF involves tricking already-authenticated users into clicking on rogue links. The rogue link asks the user to authorize a request, essentially granting the malicious attacker inside access.

The best way to guard against this is to embed a **nonce** into secured assets and refuse requests that lack them. A nonce is a semi-random number generated on the server that *marks* proper resources. The nonce is embedded as a CSRF token and must be embedded in any state-changing bits of HTML, typically forms.

If you use a templating engine that integrates tightly with Spring Boot, such as Thymeleaf, then there's no need to do anything. Thymeleaf templates will automatically add suitable CSRF-based HTML inputs to any forms rendered on the page.

Mustache, in its lightness, doesn't have such integration. However, it's possible to make Spring Security's CSRF token available by applying this inside `application.properties`:

```
spring.mustache.servlet.expose-request-attributes=true
```

With the preceding setting, a new attribute, `_csrf`, is made available to the templating engine. This makes it possible for us to update the search form as follows:

```
<form action="/search" method="post">
  <label for="value">Search:</label>
  <input type="text" name="value">
  <input type="hidden" name="{{_csrf.parameterName}}"
    value="{{_csrf.token}}">
  <button type="submit">Search</button>
</form>
```

The preceding version of the search form contains one additional hidden input. The `_csrf` token will be automatically exposed for template rendering.

Only a valid HTML template rendering on our own server will have the right `_csrf` values embedded. A rogue web page that an authenticated user gets tricked into visiting will not have these values. Since the `_csrf` value changes from request to request, there is no way for other sites to either cache or predict the values.

The other form inside index.mustache that allows us to create new videos also needs this change:

```
<form action="/new-video" method="post">
  <input type="text" name="name">
  <input type="text" name="description">
  <input type="hidden" name="{{_csrf.parameterName}}"
    value="{{_csrf.token}}">
  <button type="submit">Submit</button>
</form>
```

The preceding code shows a lightweight change that strengthens things. In fact, it's the default policy for Spring Security!

But remember how I mentioned this is an issue that requires a choice? That's because Spring Security's CSRF filter either kicks in for both our templates and our JSON-based API controllers, or we disable it for both scenarios.

To be clear, CSRF protections make perfect sense when there is a web page that we log in to. But CSRF protections aren't needed when doing API calls in a stateless scenario.

Our application is serving both scenarios, so a proper architecture would involve breaking up this application into two different applications. The web one could continue applying proper CSRF protections as shown in the preceding code.

The other application could disable CSRF protections with the following tweak to SecurityFilterChain:

```
@Bean
SecurityFilterChain configureSecurity(HttpSecurity http)
  throws Exception {
    http.authorizeHttpRequests() //
      .mvcMatchers("/login").permitAll() //
      .mvcMatchers("/", "/search").authenticated() //
      .mvcMatchers(HttpMethod.GET, "/api/**")
        .authenticated() //
        .mvcMatchers(HttpMethod.POST, "/new-video",
          "/api/**").hasRole("ADMIN") //
      .anyRequest().denyAll() //
      .and() //
      .formLogin() //
      .and() //
```

```
        .httpBasic() //
        .and() //
        .csrf().disable();
    return http.build();
}
```

The preceding piece of `SecurityConfig` is almost the same security policy we created at the end of the previous section. The only change is in the next to last line, where we have `.and().csrf().disable()`. This tiny directive tells Spring Security to completely switch off CSRF protections.

As an overall approach, let's drop `ApiController`, which was brought along from the previous chapters, and presume that it exists in a different application. Thus, there is no need to disable CSRF protections. Instead, we can roll with the changes made to `index.mustache` as shown in the previous two fragments.

With all this, we have a fistful of ways to apply path-based security. But these are not the only ways to secure our application and are not necessarily optimal for certain scenarios.

So far, we have provided a source of user data as well as created some initial access rules for our users. In the following sections, we will round things out by applying even more strategic security checks.

For example, in the next section, we'll discover method-based security practices.

Securing Spring Data methods

> **Where to find this section's code**
>
> The source code used for this portion of the chapter can be found at `https://github.com/PacktPublishing/Learning-Spring-Boot-3.0/tree/main/ch4-method-security`.

So far, we've seen tactics to apply various security provisions based on the URL of the request. But Spring Security also comes with method-level security.

While it's possible to simply apply these techniques to controller methods, service methods, and in fact, any Spring bean's method calls, this may appear to be trading one solution for another.

Method-level security specializes in providing a finer-grained ability to lock things down.

Updating our model

Before we can delve into this, we need an update to our domain model used earlier in this chapter. As a reminder, we created a `VideoEntity` class in the previous section that has an `id`, `name`, and `description` field.

To really take advantage of method-level security, we should augment this entity definition with one additional (and common) convention: adding a `username` field to represent ownership of the data:

```
@Entity
class VideoEntity {

  private @Id @GeneratedValue Long id;
  private String username;
  private String name;
  private String description;

  protected VideoEntity() {
    this(null, null, null);
  }

  VideoEntity(String username, String name, String
    description) {
      this.id = null;
      this.username = username;
      this.description = description;
      this.name = name;
    }

  // getters and setters
}
```

This updated `VideoEntity` class is pretty much the same as earlier in this chapter except for the additional `username` field. (The boilerplate getters and setters are excluded for brevity).

If we're going to dabble with ownership, it makes sense to update our set of users. Earlier in this chapter, we simply had `user` and `admin`. Let's expand the number of users to `alice` and `bob`:

```
@Bean
CommandLineRunner initUsers(UserManagementRepository
  repository) {
    return args -> {
      repository.save(new UserAccount("alice", "password",
        "ROLE_USER"));
      repository.save(new UserAccount("bob", "password",
```

```
        "ROLE_USER"));
      repository.save(new UserAccount("admin", "password",
        "ROLE_ADMIN"));
  };
}
```

Remember the `initUsers` code toward the beginning of the chapter? We are replacing that with this set of three users: `alice`, `bob`, and `admin`. Why? So we can proceed to create a security protocol where `alice` can only delete videos that she uploaded, and `bob` can only delete videos that he uploaded.

> **Alice and Bob?**
>
> Security scenarios are often described in terms of Alice and Bob, a convention used since the 1978 paper *A Method for Obtaining Digital Signatures and Public-key Cryptosystems* by *Rivest, Shamir*, and *Adleman*, the inventors of *RSA security*. See `https://en.wikipedia.org/wiki/Alice_and_Bob` for more details.

Taking ownership of data

If we're going to assign ownership to `VideoEntity` objects, then it makes sense to assign it when new entries are created. So, we should revisit the `HomeController` method that handles the `POST` requests to create new entries:

```
@PostMapping("/new-video")
public String newVideo(@ModelAttribute NewVideo newVideo,
  Authentication authentication) {
    videoService.create(newVideo,authentication.getName());
    return "redirect:/";
}
```

This `newVideo` method in `HomeController` is just like the method earlier in this chapter except it has one extra argument: `authentication`.

This is a parameter offered by Spring MVC when Spring Security is on the classpath. It automatically extracts the authentication details stored in the servlet context and populates the `Authentication` instance.

In this situation, we are extracting its `name`, a field found in the `Authentication` interface's parent interface, `java.security.Principal`, a standard type. Based on the Javadocs of `Principal`, this is *the name of the principal*.

Naturally, this requires that we update `VideoService.create` as follows:

```
public VideoEntity create(NewVideo newVideo, String
  username) {
    return repository.saveAndFlush(new VideoEntity
      (username, newVideo.name(), newVideo.description()));
}
```

The preceding updated version of `create` has two key changes:

- There's an extra argument: `username`
- `username` is passed along to the `VideoEntity` constructor we just updated

These changes will make every newly entered video automatically associate with the currently logged-in user.

Adding a delete button

We talked about providing the ability to delete videos, but locking down that feature only to the owner of the video. Let's tackle this step-by-step, starting with rendering each video along with a *DELETE* button inside `index.mustache` as follows:

```
{{#videos}}
    <li>
        {{name}}
        <form action="/delete/videos/{{id}}" method="post">
            <input type="hidden"
                name="{{_csrf.parameterName}}"
                value="{{_csrf.token}}">
            <button type="submit">Delete</button>
        </form>
    </li>
{{/videos}}
```

The preceding fragment of `index.mustache` can be described as follows:

- The `{{#videos}}` tag tells Mustache to iterate over the array of the videos attribute
- It will render an HTML line for every instance found in the database
- `{{name}}` will render the name field

- The `<form>` entry will create an HTML form with the `{{id}}` field used to build a link to `/delete/videos/{{id}}`

It's important to understand that this form has a hidden `_csrf` input, as we discussed earlier in this chapter. Since this is HTML and not REST, we are using POST and not DELETE as our HTTP verb of choice.

Now, we need to add a method to `HomeController` that responds to POST `/delete/videos/{{id}}` calls as follows:

```
@PostMapping("/delete/videos/{videoId}")
public String deleteVideo(@PathVariable Long videoId) {
  videoService.delete(videoId);
  return "redirect:/";
}
```

This method can be described as follows:

- `@PostMapping` signals that this method will respond to POST on URL `/delete/videos/{videoId}`
- The `@PathVariable` extracts the `videoId` argument based on name matching
- Using the `videoId` field, we pass it along to `VideoService`
- If things work out, the method returns `"redirect:/"`, a Spring MVC directive to issue an **HTTP 302 Found**, a soft redirect that bounces the user back to GET `/`

Next, we need to create the `delete()` method inside `VideoService` as follows:

```
public void delete(Long videoId) {
  repository.findById(videoId) //
    .map(videoEntity -> {
      repository.delete(videoEntity);
      return true;
    }) //
    .orElseThrow(() -> new RuntimeException("No video at "
                                            + videoId));
}
```

The preceding method can be described as follows:

- `videoId` is the primary key for the video to be deleted.
- We first use the repository's `findById` method to look up the entity.

- Spring Data JPA returns `Optional`, which we can map over to get the `VideoEntity` object.

- Using the `VideoEntity` object, we can then perform the `delete(entity)` method. Because `delete()` has a return type of `void`, we have to return `true` to comply with `Optional.map`'s demand for a return value.

- If `Optional` turns out to be empty, we will instead throw `RuntimeException`.

Locking down access to the owner of the data

We're almost there. So far, in this chapter we've written code that transform's a video's `id` field into a `delete` operation. But this section is about restricting access with Spring Data methods. While Spring Data JPA's `JpaRepository` interface has a handful of delete operations, we must extend this definition inside `VideoRepository` if we wish to apply security controls as follows:

```
@PreAuthorize("#entity.username == authentication.name")
@Override
void delete(VideoEntity entity);
```

The preceding change to Spring Data JPA's built-in `delete(VideoEntity)` method can be described as follows:

- `@Override`: This annotation will ensure that we don't alter the name of the method or any other aspect of the method signature

- `@PreAuthorize`: This is Spring Security's method-based annotation that allows us to write customized security checks

- `#entity.username`: This de-references the entity argument in the first parameter and then looks up the username parameter using Java bean properties

- `authentication.name`: A Spring Security argument to access the current security context's authentication object and look up the principal's name

By comparing the `username` field of `VideoEntity` against the current user's `name`, we can confine this method to *only* working when *a user attempts to delete one of their own videos*.

Enabling method-level security

Now, none of this will work if we don't enable method-level security. We must circle back to the `SecurityConfig` class we created in the first parts of this chapter and add the following annotation:

```
@Configuration
@EnableMethodSecurity
public class SecurityConfig {
```

```
    // prior security configuration details
}
```

In the preceding code, `@EnableMethodSecurity` is Spring Security's annotation to activate method-based security.

> **Note**
>
> Perhaps you've heard of `@EnableGlobalMethodSecurity`? If so, be aware that this is being phased out in favor of `@EnableMethodSecurity` (used in the preceding code). For starters, the newer `@EnableMethodSecurity` activates Spring Security's more powerful `@PreAuthorize` annotation (and its cousins) by default, while leaving the antiquated `@Secured` annotation disabled as well as the somewhat limited **JSR-250** (`@RolesAllowed`) annotations. In addition to this, `@EnableMethodSecurity` leverages Spring Security's simplified `AuthorizationManager` API instead of the more complex metadata sources, config attributes, decision managers, and voters.

This sets things up!

Displaying user details on the site

Just a little something to top things off, we are now going to learn about displaying user-based security details along with a logout button.

For this, we first must update `HomeController` as follows:

```
@GetMapping
public String index(Model model,
   Authentication authentication) {
      model.addAttribute("videos", videoService.getVideos());
      model.addAttribute("authentication", authentication);
      return "index";
   }
```

The preceding controller method is the same as earlier in this chapter but with a few changes in `Authentication`. Like the `delete()` controller method created earlier in this section, we are also tapping into this provided value by storing the details in an extra model attribute.

This mechanism allows us to provide authentication details for the current user to the template. We'll take advantage of it by updating index.mustache toward the top as follows:

```
<h3>User Profile</h3>
<ul>
    <li>Username: {{authentication.name}}</li>
    <li>Authorities: {{authentication.authorities}}</li>
</ul>

<form action="/logout" method="post">
    <input type="hidden" name="{{_csrf.parameterName}}"
      value="{{_csrf.token}}">
    <button type="submit">Logout</button>
</form>
```

The preceding fragment of HTML can be described as follows:

- The username is shown by using {{authentication.name}}.

- The authorities are shown by using {{authentication.authorities}}.

- A logout button is provided inside an HTML form. The action to log out with Spring Security is to POST against /logout. It's vital to know that even logging out requires providing the _csrf token if CSRF hasn't been disabled!

> **Important**
>
> If it's not clear, every single HTML form in index.mustache (and any other templates you've decided to add) must have the _csrf token as a hidden input. Spring Boot doesn't have the same level of integration with Mustache's template engine that it has with *Thymeleaf*. If you use Thymeleaf, while experiencing a steeper learning curve, it will automatically add these hidden inputs, freeing you up from having to remember them!

Earlier in this chapter, we preloaded some VideoEntity data inside the VideoService. Let's update this based on alice and bob as follows:

```
@PostConstruct
void initDatabase() {
  repository.save(new VideoEntity("alice", "Need HELP with
    your SPRING BOOT 3 App?",
    "SPRING BOOT 3 will only speed things up and make it
```

```
      super SIMPLE to serve templates and raw data."));
  repository.save(new VideoEntity("alice", "Don't do THIS
    to your own CODE!",
    "As a pro developer, never ever EVER do this to your
      code. Because you'll ultimately be doing it to
        YOURSELF!"));
  repository.save(new VideoEntity("bob", "SECRETS to fix
    BROKEN CODE!",
    "Discover ways to not only debug your code, but to
      regain your confidence and get back in the game as a
        software developer."));
}
```

In the preceding method, the bottom of `VideoService` can be described as follows:

- `@PostConstruct`: A standard Jakarta EE annotation that signals for this method to be run after the application has started

- `repository`: The `VideoRepository` field is used to load two videos for `alice` and one for `bob`

With all this in place, we can restart the application and take it for a spin!

If we visit `localhost:8080`, we are instantly bounced over to Spring Security's pre-built login page:

Figure 4.2 – Logging in as alice

If we log in as `alice`, we are presented with the following message at the top of the page:

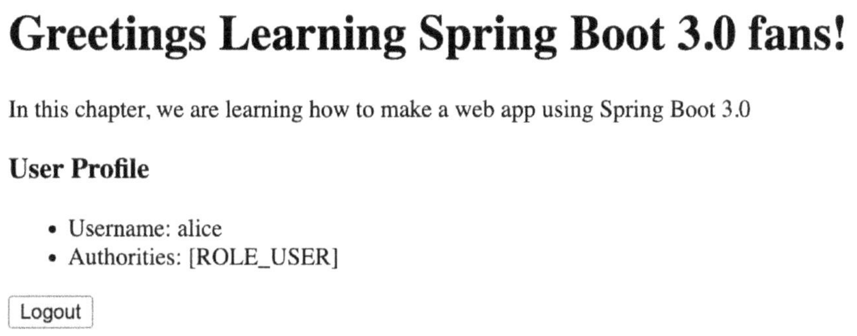

Greetings Learning Spring Boot 3.0 fans!

In this chapter, we are learning how to make a web app using Spring Boot 3.0

User Profile

- Username: alice
- Authorities: [ROLE_USER]

Logout

Figure 4.3 – The index template rendering user authentication details

The preceding page shows the extra security details. It has both the **Username** field as well as the user's assigned **Authorities**. Finally, there is a **Logout** button.

> **Important!**
>
> Do NOT put the user's password on the page! In fact, it's probably best to not put the list of authorities either. *Figure 4.3* only illustrates the amount of information given to the template to allow sophisticated options. While Mustache is a bit limited in its logic-less nature, Thymeleaf actually has Spring Security extensions that let you optionally render sections and perform security checks. Why aren't we covering this in this book? Because this book is titled *Learning Spring Boot 3.0*, not *Learning Thymeleaf*. But it's only fair to let you know that you have options.

What happens if, as `alice`, we clicked on **DELETE** in the last video (owned by Bob)? Alice will see the following **403** response:

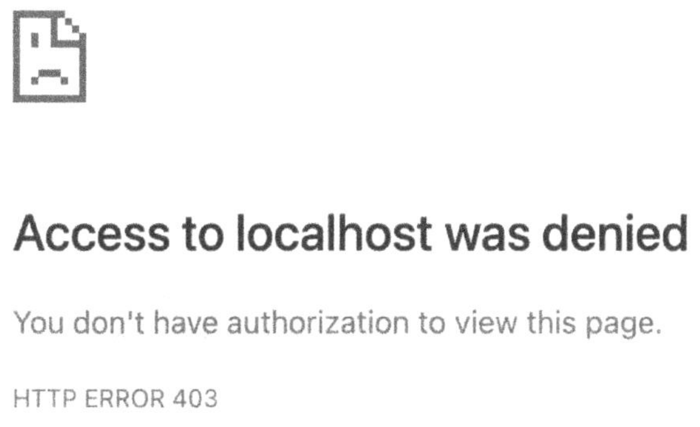

Access to localhost was denied

You don't have authorization to view this page.

HTTP ERROR 403

Figure 4.4 – When Spring Security denies access, you get served a 403 forbidden page

Hit the **Back** button on the browser and try to delete one of your own videos, and things will work just fine!

> **Bonus**
>
> For a higher-level, more visual perspective on Spring Security, check out this video at `https://springbootlearning.com/security`.

With the method-based controls we have covered in this section, we are able to implement fine-grained access. We are controlling things not just at the URL level, but also at the object level. But this comes at a price.

Someone has to manage all these users and their roles. Don't be surprised if you have to set up a security ops team simply to manage the users. User management can be quite tedious.

This is what drives many teams to consider outsourcing user management entirely to alternative tools, such as Facebook, Twitter, GitHub, Google, or Okta.

We'll explore how to consider using Google as an identity provider in the next section.

Leveraging Google to authenticate users

Do you dread the thought of managing users and their passwords? Many security teams buy large products to deal with all this. Teams even invest in tools to simply push password resets directly to users, to reduce call volume.

Long story short, user management is a major effort not to be taken lightly; hence, many teams turn to **OAuth**. Described as "*an open standard for access delegation*" (`https://en.wikipedia.org/wiki/OAuth`), OAuth provides a way to outsource user management almost entirely.

OAuth arose as social media applications emerged. A user of a third-party Twitter app used to store their password directly in the app. Not only was this inconvenient when users wanted to change their password, but it was a major security risk!

OAuth lets the application move away from this by instead reaching out to the social media site directly. The user logs in with the social media site, and the site hands back a special token to the app containing well-defined privileges known as **scopes**. This empowers the application to interact on the user's behalf.

So, it should come as no surprise that literally every social media service has OAuth support. We can reach out to any application, be it Twitter, Facebook, GitHub, Google, and more, and simply let our users log in there. In fact, some apps support ALL these sites, making it possible to easily let users into your application.

This is not our only choice. We can also establish our own OAuth user service. While this may sound like circling back after pitching the reason to *not* run our own user management service, there are pros and cons to these various choices.

Pros of using OAuth

- If you use Facebook or Twitter, anyone who already has a presence there can access your app. Given the popularity of these platforms, this can be an easy way to allow your users access.

- If you choose GitHub, then your user base should be heavily oriented toward developers. Not every developer has a GitHub account, but a large number do. This can be a boon for apps that are tilted in this direction.

- If you choose Google, another vast collection of users will be available to you.

- If you choose Okta, a commercial system you can configure and use, then you have 100% control. Your users don't have to exist on any social media platform and they don't have to be developer oriented. You have complete control while still being able to outsource the hard parts of user management.

Cons of using OAuth

- If someone doesn't have a presence on your preferred social media network, they must either open an account or choose not to access it. You must *also* support managing your own users, which we're trying to get away from.

- If you choose Facebook, Twitter, Google, GitHub, or any other social media site, you are confined to *their* scopes. You don't get to define your own. If the goal is to simply access and leverage user information, this is probably fine. However, if you wish to have managers, board officers, admins, DBAs, and other assorted roles, this won't suffice.

- If your application doesn't take full advantage of GitHub (for example, `https://gitter.im`), then using GitHub probably isn't the way to go.

If you need full control of scopes, Okta is the way to go. To top it off, Okta's development team maintains full integration with Spring Security.

Assuming we've made our assessment and made a choice, it's time to configure Spring Security to hook in.

We could pick any of the mentioned services to illustrate this, but for the rest of this book, let's go with Google.

Creating a Google OAuth 2.0 application

Before we can take any steps toward authenticating with Google, we must first create an application with Google. To be clear, this means we will register some details on their dashboard, extract credentials, and plug these credentials into our Spring Boot application, granting our users access to Google data.

To do this, go through the following steps:

1. Go to Google Cloud's Dashboard (`https://console.cloud.google.com/home/dashboard`).

2. Click on the drop-down right next to **Select a project** in the top-left corner. Hit **NEW PROJECT** on the popup and accept the default values.

3. Select your new project so it's showing in the drop-down at the top.

4. On the left-hand panel on the Google Cloud Dashboard, scroll down and hover over **APIs and services**. On the pop-up menu, click on **Enabled APIs and services**.

5. On the list toward the bottom of the page, look for **YouTube Data API v3**. Click on it and hit **Enable API**. This will grant our application access to YouTube's latest data API.

6. Back at your newly created application's dashboard, look on the left-hand panel, and select **Credentials**.

7. Click on + **CREATE CREDENTIALS**. On the pop-up menu, select **OAuth Client**.

8. For application type, select **Web Application**.

9. In the **Name** entry, give the application a name.

10. In the **Authorized redirect URIs**, enter `http://localhost:8080/login/oauth2/code/google`.

11. Once these credentials are created, capture the **Client ID** and **Client secret** shown in the top-right corner. We'll later plug them into our Spring Boot application.

12. Go back to the left-hand column from earlier (**APIs and services** at `https://console.cloud.google.com/apis/dashboard`) and click on the **OAuth consent screen**.

13. Underneath test users, create an entry for each email address you wish to log in under later when we build the Spring Boot application.

These steps may seem tedious, but all modern-day OAuth applications on any platform, Google or otherwise, will require some amount of the following:

- Application definition
- Approved platform APIs
- Supported users
- Callbacks to our own application

It's really just a matter of digging through the console and finding where the settings are plugged in.

> **Note**
>
> At this stage, our Google application is considered to be in test mode. This means that we are the only ones who can access it. We'll be able to run the code locally on our machine and shake it out. No one else can access anything unless we publish it.

Adding OAuth Client to a Spring Boot project

> **Where to find this section's code**
>
> The source code for this portion of the chapter can be found at `https://github.com/PacktPublishing/Learning-Spring-Boot-3.0/tree/main/ch4-oauth`.

Earlier in this chapter, we took the code from the previous chapter, *Querying for Data with Spring Boot*, and added Spring Security.

In this section, we actually need to start afresh:

1. Go and visit **Spring Initializr** at `https://start.spring.io`.
2. Select or enter the following details:

    -   **Project: Maven Project**
    -   **Language: Java**
    -   **Spring Boot: 3.0.0**
    -   **Group**: `com.springbootlearning.learningspringboot3`
    -   **Artifact**: `ch4-oauth`
    -   **Name**: `Chapter 4 (OAuth)`
    -   **Description**: `Securing an Application with Spring Boot and OAuth 2.0`
    -   **Package name**: `com.springbootlearning.learningspringboot3`
    -   **Packaging: Jar**
    -   **Java: 17**

3. Click on **ADD DEPENDENCIES**. Then, select the following:

    -   **OAuth 2 Client**
    -   **Spring Web**
    -   **Spring Reactive Web**
    -   **Mustache**

4.   Click on the **GENERATE** button at the bottom of the screen.

5.   Import the code into your favorite IDE.

First of all, what are we doing with both Spring Web and Spring Reactive Web? Spring Web is how we built a servlet-based web application using Spring MVC. But another feature we are going to use is Google's Oauth API, which leverages *WebClient*, Spring's next-generation HTTP client. It's found in Spring Reactive *WebFlux*.

When Spring Boot sees both Spring Web and Spring Reactive Web on the classpath, it will default to a standard embedded Apache Tomcat and run a standard servlet container.

With all this in place, we can start building our OAuth web application!

The Spring Initializr creates an `application.properties` file as part of the generated project. Due to the repetitive nature of some of the properties we must access, we will switch to the YAML-based variant. Simply rename that file as `application.yaml`.

Inside that file, add the following entry:

```yaml
spring:
  security:
    oauth2:
      client:
        registration:
          google:
            clientId: **your Google Client ID**
            clientSecret: **your Google Client secret**


            scope: openid,profile,email,
                   https://www.googleapis.com/auth/youtube
```

In the preceding code, `application.yaml` lets us enter properties in a hierarchical fashion. This is most helpful when we have to enter multiple properties at the same sublevel. In this case, we have several entries at `spring.security.oauth2.client.registration.google`.

> **Warning**
>
> OAuth2, due to its incredible flexibility, comes with many settings. This is necessary due to the flow of users coming to our app, being forwarded over to the other platform for authentication, and then flowing back to our application. To ease setup, Spring Security has added `CommonOAuth2Provider` with pre-baked settings for Google, GitHub, Facebook, and Okta. We only have to plug in `clientId` and `clientSecret`. Technically, that's enough to authenticate with Google. But since we plan to leverage the YouTube Data API, we have added a scope setting, which we'll discuss later.

The web we are building, which speaks to Google, can be described as an OAuth2-authorized client. This is represented in Spring Security OAuth2 using OAuth2AuthorizedClient. To facilitate the flow between our application and Google, Spring Boot autoconfigures ClientRegistrationRepository as well as OAuth2AuthorizedClientRepository.

These are the classes that parse the clientId and clientSecret properties in the preceding application.yaml fragment. An additional reason we need these repositories is that OAuth2 supports *working with more than one OAuth2 provider*.

Surely you've seen certain websites that support letting you log in using multiple options including Facebook, Twitter, Google, and perhaps even Apple?

Hence, we need some functionality to broker all these requests. To do this, we need to create a SecurityConfig class and add the following bean definition:

```
@Configuration
public class SecurityConfig {

    @Bean
    public OAuth2AuthorizedClientManager clientManager(
      ClientRegistrationRepository clientRegRepo,
        OAuth2AuthorizedClientRepository authClientRepo) {

      OAuth2AuthorizedClientProvider clientProvider =
        OAuth2AuthorizedClientProviderBuilder.builder()
          .authorizationCode()
          .refreshToken()
          .clientCredentials()
          .password()
          .build();

      DefaultOAuth2AuthorizedClientManager clientManager =
        new DefaultOAuth2AuthorizedClientManager(
          clientRegRepo, authClientRepo);
      clientManager
        .setAuthorizedClientProvider(clientProvider);

      return clientManager;
    }
}
```

The preceding `clientManager()` bean definition will request the two autoconfigured Oauth2 beans mentioned earlier and blend them together into `DefaultOAuth2AuthorizedClientManager`. This bean will do the legwork of pulling the necessary properties from `application.yaml` and using them in the context of an incoming servlet request.

> **Tip**
>
> The flow of OAuth2 can seem a bit daunting. This is why you may wish to read more about it at `https://oauth.net/2/`, the official site for the OAuth 2.0 spec. It's also important to appreciate that yes, this is a boilerplate, but once completed, we'll have a valuable resource available for our Spring Boot application.

Believe it or not, we're close to taking advantage of Google as our OAuth 2 platform of choice. When Spring Security OAuth2 is put on the classpath, Spring Boot has autoconfiguration policies that will automatically lock down our application. But this time, instead of creating a random password for a fixed username, the OAuth 2 beans mentioned earlier are combined with `OAuth2AuthorizationClientManager`.

But we want to go one step further. We want to actually invoke Google's YouTube Data API, which we'll cover in the next section.

Invoking an OAuth2 API remotely

To hook OAuth 2 support into an HTTP remote service invoker such as `WebClient`, start by creating a class named `YouTubeConfig` and add the following bean definitions:

```
@Configuration
public class YouTubeConfig {

  static String YOUTUBE_V3_API = //
    "https://www.googleapis.com/youtube/v3";


  @Bean
  WebClient webClient(OAuth2AuthorizedClientManager
                       clientManager) {

  ServletOAuth2AuthorizedClientExchangeFilterFunction
    oauth2 = //
     new
       ServletOAuth2AuthorizedClientExchangeFilterFunction(
        clientManager);
```

```
    oauth2.setDefaultClientRegistrationId("google");

    return WebClient.builder() //
      .baseUrl(YOUTUBE_V3_API) //
      .apply(oauth2.oauth2Configuration()) //
      .build();
  }
}
```

The preceding bean definition creates a new `WebClient`, Spring's more recent HTTP remote service invoker. While the venerable `RestTemplate` isn't going anywhere, `WebClient` offers many new-and-improved ways to interact with remote HTTP services. One of the biggest improvements is its fluent API along with fully realized support for reactive services. While we aren't using them in this chapter, this is something we'll touch on later in this book.

In the preceding code, not only is this `WebClient` pointed at Google's YouTube v3 API, but it also registers an **exchange filter function** using our previously created `OAuth2AuthorizedClientManager` as a way to give it OAuth2 power.

> **Note**
>
> An exchange filter function is a concept not found in servlets and Spring MVC, but instead in Spring WebFlux's reactive paradigm. It's very similar to a classic *servlet filter* in that every request that passes through the WebClient will invoke this function. This will ensure that the current user is logged in to Google and has the right authorization.

Another reason we may wish to use Spring WebFlux's `WebClient`, despite having a servlet-based application, is to leverage one of Spring Framework's most recent additions: **HTTP client proxies**.

The idea of HTTP client proxies is to capture all the details needed to interact with a remote service in an interface definition and let Spring Framework, under the hood, marshal the request and response.

We can capture one such exchange by creating an interface named *YouTube*, as follows:

```
interface YouTube {

    @GetExchange("/search?part=snippet&type=video")
    SearchListResponse channelVideos( //
      @RequestParam String channelId, //
      @RequestParam int maxResults, //
      @RequestParam Sort order);
```

```
enum Sort {
  DATE("date"), //
  VIEW_COUNT("viewCount"), //
  TITLE("title"), //
  RATING("rating");

  private final String type;

  Sort(String type) {
    this.type = type;
  }
}
}
```

The preceding interface has a single method: `channelVideos`. In truth, the name of this method doesn't matter, because it's the `@GetExchange` method that matters. In *Chapter 2, Creating a Web Application with Spring Boot*, we saw how to use `@GetMapping` to link HTTP GET operations with Spring MVC controller methods.

For HTTP remoting, the counterpart annotation is `@GetExchange`. This tells Spring Framework to remotely invoke `/search?part=snippet&type=video` using an HTTP GET call.

If it's not obvious, the path in the `@GetExchange` call is appended to the base URL configured earlier, `https://www.googleapis.com/youtube/v3`, forming a complete URL to access this API.

In addition to specifying the HTTP verb along with the URL, this method has three inputs: `channelId`, `maxResults`, and `order`. The `@RequestParam` annotation indicates that these parameters are to be added to the URL as query parameters in the form of `?channelId=<value>&maxResults=<value>&order=<value>`. To top things off, the `order` parameter is constrained to what the API considers acceptable values using a Java enum, `Sort`.

The names of the query parameters are lifted from the method's argument names. While it's possible to override these using the `@RequestParam` annotation, I find it easier to simply set each argument's name to match the API.

For this particular API, there is no HTTP request body (as can be seen on the APIs documentation at `https://developers.google.com/youtube/v3/docs/search/list`). The whole request is contained in the URL.

But if you did need to send over some data to a different API, perhaps one expecting an HTTP POST, then you can use `@PostExchange`. You would provide the data as another method argument and apply `@RequestBody` so that Spring Framework knows to ask **Jackson** to serialize the provided data into JSON.

The response from this API is a JSON document shown in detail further down in the **Response** section of the **Search** function in the preceding list, as follows:

```
{
    "kind": "youtube#searchListResponse",
    "etag": etag,
    "nextPageToken": string,
    "prevPageToken": string,
    "regionCode": string,
    "pageInfo": {
        "totalResults": integer,
        "resultsPerPage": integer
    },
    "items": [
        search Resource
    ]
}
```

Java 17 records really shine while capturing the preceding API response. We need some barebone Java objects that are mostly data-oriented, and we need them fast. So, create a record called `SearchListResponse` as follows:

```
record SearchListResponse(String kind, String etag, String
  nextPageToken, String prevPageToken, PageInfo pageInfo,
    SearchResult[] items) {
}
```

We can include all the fields we want and leave out the ones we don't care about. In the preceding code, most of the fields are plain old Java strings, but the last two, `PageInfo` and `SearchResult`, are not.

So, create some more Java records, putting each in its own file:

```
record PageInfo(Integer totalResults, Integer
  resultsPerPage) {
}
record SearchResult(String kind, String etag, SearchId id,
  SearchSnippet snippet) {
}
```

The process to create the preceding types is to simply walk through each nested type on Google's YouTube API documentation, and capture the fields as shown. They'll describe them as strings, integers, nested types, or links to other types. For each subtype that has its own section, simply create another record!

> **Tip**
> The name of the record's *type* doesn't matter. The critical part is ensuring the name of the *field* matches the names in the JSON structure being passed back.

With what we've captured so far in records, we now need to create `SearchId` as well as `SearchSnippet`, again, each in its own file:

```
record SearchId(String kind, String videoId, String
   channelId, String playlistId) {
}
record SearchSnippet(String publishedAt, String channelId,
   String title, String description,
     Map<String, SearchThumbnail> thumbnails, String
       channelTitle) {
}
```

These record types are almost complete, given that they are almost all built-in Java types. The only one missing is `SearchThumbnail`. If we read YouTube's API reference docs, we can easily wrap things up with this record definition:

```
record SearchThumbnail(String url, Integer width, Integer
   height) {
}
```

This last record type is just a string and a couple of integers, so we're done!

But not quite. We've invested loads of time configuring OAuth 2 and a remoting HTTP service to talk to YouTube. The cherry on top is building the web layer of our app, shown in the next section.

Creating an OAuth2-powered web app

Now, it's time to create a web controller to start rendering things:

```
@Controller
public class HomeController {
```

```
  private final YouTube youTube;

  public HomeController(YouTube youTube) {
    this.youTube = youTube;
  }

  @GetMapping
  String index(Model model) {
    model.addAttribute("channelVideos", //
      youTube.channelVideos("UCjukbYOd6pjrMpNMFAOKYyw",
        10, YouTube.Sort.VIEW_COUNT));
    return "index";
  }
}
```

The preceding web controller has some key points:

- `@Controller` indicates that this is a template-based web controller. Each web method returns the name of a template to render.

- We are injecting the YouTube service through **constructor injection**, a concept touched upon back in *Chapter 2, Creating a Web Application with Spring Boot.*

- The `index` method has a Spring MVC Model object, where we create a `channelVideos` attribute. It invokes our YouTube service's `channelVideos` method with a channel ID, a page size of 10, and uses view counts as the way to sort search results.

- The name of the template to render is `index`.

Since we are using Mustache as our templating engine of choice, the name of the template expands to `src/main/resources/templates/index.mustache`. To define it, we can start coding some pretty simple HTML 5 as follows:

```
<!doctype html>
<html lang="en">
<head>
    <link href="style.css" rel="stylesheet"
    type="text/css"/>
</head>
<body>
<h1>Greetings Learning Spring Boot 3.0 fans!</h1>
```

```html
<p>
    In this section, we are learning how to make
    a web app using Spring Boot 3.0 + OAuth 2.0
</p>

<h2>Your Videos</h2>
<table>
    <thead>
    <tr>
        <td>Id</td>
        <td>Published</td>
        <td>Thumbnail</td>
        <td>Title</td>
        <td>Description</td>
    </tr>
    </thead>
    <tbody>
    {{#channelVideos.items}}
        <tr>
            <td>{{id.videoId}}</td>
            <td>{{snippet.publishedAt}}</td>
            <td>
                <a href="https://www.youtube.com/watch?v=
                {{id.videoId}}" target="_blank">
                <img src="{{snippet.thumbnail.url}}"
                alt="thumbnail"/>
                </a>
            </td>
            <td>{{snippet.title}}</td>
            <td>{{snippet.shortDescription}}</td>
        </tr>
    {{/channelVideos.items}}
    </tbody>
</table>
</body>
```

We aren't going to explore every facet of HTML 5 in the preceding code. But some of the key bits of Mustache are as follows:

- Mustache directives are wrapped in double curly braces, whether it's to iterate over an array (`{{#channelVideos.items}}`) or a single field (`{{id.videoId}}`).

- A Mustache directive that starts with a pound sign (#) is a signal to iterate, generating a copy of HTML for every entry. Because the `SearchListResponse` items fields are an array of `SearchResult` entries, the HTML inside that tag is repeated for each entry.

- The thumbnail field inside `SearchSnippet` actually has multiple entries for each video. As Mustache is a logic-less engine, we need to augment that record definition with some extra methods to support our templating needs.

Adding a way to pick the right thumbnail and also curtailing the description field to less than a hundred characters can be implemented by updating the `SearchSnippet` record as follows:

```
record SearchSnippet(String publishedAt, String channelId,
  String title, String description,
    Map<String, SearchThumbnail> thumbnails, String
      channelTitle) {

  String shortDescription() {
    if (this.description.length() <= 100) {
      return this.description;
    }
    return this.description.substring(0, 100);
  }

  SearchThumbnail thumbnail() {
    return this.thumbnails.entrySet().stream()
      .filter(entry -> entry.getKey().equals("default"))
      .findFirst()
      .map(Map.Entry::getValue)
      .orElse(null);
  }
}
```

From the preceding code, we can see that the following will occur:

- The `shortDescription` method will either return the `description` field directly or a 100-character substring

- The `thumbnail` method will iterate over the thumbnail map entries, find the one named `default`, and return it

As a tasty finish, let's apply CSS so our table comes out nicely polished. Create `src/main/resources/static/style.css`:

```css
table {
    table-layout: fixed;
    width: 100%;
    border-collapse: collapse;
    border: 3px solid #039E44;
}

thead th:nth-child(1) {
    width: 30%;
}

thead th:nth-child(2) {
    width: 20%;
}

thead th:nth-child(3) {
    width: 15%;
}

thead th:nth-child(4) {
    width: 35%;
}

th, td {
    padding: 20px;
}
```

As per the preceding code, Spring MVC will serve up static resources found in `src/main/resources/static` automatically. With all this in place, let's launch the application! Visit `localhost:8080` and we should automatically get forwarded to Google's login page:

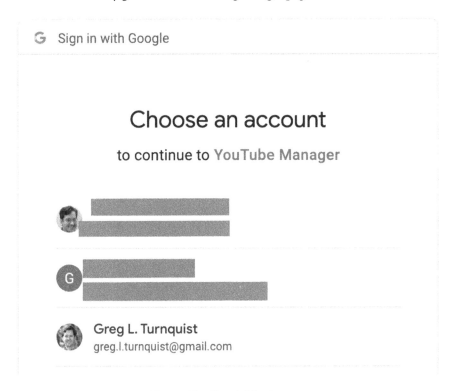

Figure 4.5 – Google's login page

This login page will show any Google accounts you have used (I have several!). What's important is to pick an account that you registered earlier on the Google Cloud Dashboard with the application. Otherwise, it won't work!

Now if we had accepted `CommonOAuth2Provider`'s standard Google scope list, then all we'd be asking of Google is for user account details, such as an email address. Then, we would have been redirected back to our own web app.

But since we customized the scope property in order to tap into the YouTube API, another prompt will pop up, asking us to select a specific YouTube channel (If you don't have one, you'll have to declare one, even if you don't upload anything!).

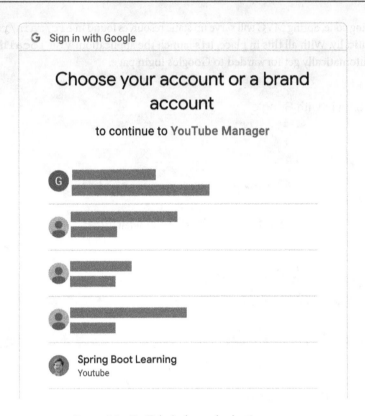

Figure 4.6 – YouTube's channel selection page

Pick your channel (Yes, I have several!). From here, we'll get redirected back to our Spring Boot Mustache template, as follows:

Greetings Learning Spring Boot 3.0 fans!

In this section, we are learning how to make a web app using Spring Boot 3.0 + OAuth 2.0

Your Videos

Id	Published	Thumbnail	Title	Description
KNcemsIbKcw	2021-03-12T12:00:12Z		SPRING BOOT and SPRING NATIVE	SPRING BOOT and SPRING NATIVE: Creating apps that start fast has never been easier! With the Sprin
v7xX6FxBhLs	2021-07-22T17:55:16Z		QUERY for DATA with SPRING BOOT	No web app is complete without fetching data. Find out why Spring Data is THE WAY to not only write
DD_Q4jGJsZ0	2021-04-16T18:42:36Z		5 WAYS to go to PRODUCTION with SPRING BOOT (ft. Josh Long)!	5 WAYS to go to PRODUCTION with SPRING BOOT: How are you going to get your app out to your users?

Figure 4.7 – Spring Boot YouTube-flavored template

Just like that, we have YouTube data on display on our own web page! The thumbnails even have hyperlinks, allowing you to open a new browser tab and watch the videos.

> **Tip**
>
> The preceding code is fetching data from my YouTube channel, showing the most popular videos (*Figure 4.7* has been trimmed to fit this book). However, you can plug in any channel ID (not vanity channel URL) and get a readout. For example, check out my friend Dan Vega's channel, a Spring developer advocate who has made multiple Spring videos, by typing in `youTube.channelVideos("UCc98QQw1D-y38wg6mO3w4MQ", 10, YouTube. Sort.VIEW_COUNT)`.

By using OAuth2, we have successfully built a system where we can offload user management to a third-party service. This greatly reduces our own risk when it comes to user management by offloading it to Google (or whatever OAuth2 service we choose).

In fact, this is a prime reason to leverage a third-party service. There are many start-ups and businesses that offer keen services, requiring little more than a user id and email address to identify users.

Ever notice how some websites *even offer* to let you log in through *multiple* external services? That's because we can actually define entries on multiple platforms, add all their entries to `application.yaml`, and then coordinate them all.

I'll leave it to you as an exercise to tinker with your application to support multiple external services.

> **Bonus**
>
> Feel free to join me for a hands-on live stream, where we literally create a Google OAuth2 client, register it with Spring Boot 3, and serve up YouTube data at `https://springbootlearning.com/oauth2`!

Summary

Over the course of this chapter, we have learned how to secure a Spring MVC application. We plugged in custom users, applied path-based controls, and even added method-level fine-grained controls using Spring Security. We topped things off by outsourcing user management to the lofty Google using Spring Security's OAuth2 integration. We took advantage of this by grabbing hold of some YouTube data and serving up video links.

This chapter may seem long, but in truth, security is a complex beast. Hopefully, with the various tactics shown in this chapter, you'll have some solid ideas on what to do when it's time to secure your own applications.

In the next chapter, *Testing with Spring Boot*, we'll explore how to ensure our code is rock solid with various testing mechanisms.

5

Testing with Spring Boot

In the previous chapter, we learned how to secure an application through various tactics, including path-based and method-based rules. We even learned how to delegate to an external system such as Google to offload risk for user management.

In this chapter, we'll learn about **testing** in Spring Boot. Testing is a multi-faceted approach. It also isn't something that you ever really finish. That's because every time we add new features, we should add corresponding test cases to capture requirements and verify they are met. It's always possible to uncover corner cases we didn't think of. And as our application evolves, we have to update and upgrade our testing methods.

Testing is a philosophy that, when embraced, enables us to grow our confidence in the software we build. In turn, we can carry this confidence to our customers and clients, demonstrating quality.

The point of this chapter is to introduce a wide range of testing tactics and their various tradeoffs. Not so we can ensure this book's sample code is well tested, but so that you can learn how to better test your projects and know what tactics to use and when!

In this chapter, we'll cover the following topics:

- Adding JUnit and other test toolkits to our application
- Creating domain-based test cases
- Testing web controllers using MockMVC
- Testing data repositories with mocks
- Testing data repositories with embedded databases
- Testing data repositories using containerized databases
- Testing security policies with Spring Security Test

> **Where to find this chapter's code**
>
> The source code for this chapter can be found at `https://github.com/PacktPublishing/Learning-Spring-Boot-3.0/tree/main/ch5`.

Adding JUnit 5 to the application

The first step in writing test cases is adding the necessary test components. The most widely accepted test tool is **JUnit**. **JUnit 5** is the latest version with deep integrations with Spring Framework and Spring Boot. (See `https://springbootlearning.com/junit-history` for more history of the origins of JUnit.)

What does it take to add JUnit to our application?

Nothing at all.

That's right. Remember how we used Spring Initialzr (`start.spring.io`) in previous chapters to roll out our new project (or augment an existing one)? One of the dependencies that was *automatically added at the bottom* is this:

```
<dependency>
        <groupId>org.springframework.boot</groupId>
        <artifactId>spring-boot-starter-test</artifactId>
        <scope>test</scope>
</dependency>
```

This test-scoped Spring Boot starter contains a fully loaded set of handy dependencies, including the following:

- **Spring Boot Test**: Spring Boot-oriented test utilities
- **JSONPath**: The query language for JSON documents
- **AssertJ**: Fluent API for asserting results
- **Hamcrest**: Library of matchers
- **JUnit 5**: Cornerstone library for writing test cases
- **Mockito**: Mocking framework for building test cases
- **JSONassert**: Library of assertions aimed at JSON documents
- **Spring Test**: Spring Framework's test utilities
- **XMLUnit**: Toolkit for verifying XML documents

If you've never heard of it, **mocking** is a form of testing where instead of checking the results, we verify the methods invoked. We'll see how to use this further in this chapter.

Simply put, *ALL* of these toolkits are already at our fingertips, ready to write tests. Why?

So that we don't have to pick and choose and waste time. There's no need to hunt down the *right* testing kit, and we don't even have to choose to test. Spring Initialzr, the champion of building Spring Boot projects, adds all of them without even requiring that we remember to add them.

Testing is that important to the Spring team.

We aren't going to necessarily use every single one of these toolkits in this chapter, but we'll get a nice cross-section of functionality. And by the end of this chapter, you should have a better perspective on what these toolkits offer.

Creating tests for your domain objects

Earlier, we mentioned that testing is a multi-faceted approach. One of the most critical things in any system is its domain types. Testing them is vital. Essentially, anything publicly visible to users is a candidate for writing test cases.

So, let's start by writing some test cases around the `VideoEntity` domain object we defined back in *Chapter 3, Querying for Data with Spring Boot*:

```java
public class CoreDomainTest {

  @Test
  void newVideoEntityShouldHaveNullId() {
    VideoEntity entity = new VideoEntity("alice",
      "title", "description");
    assertThat(entity.getId()).isNull();
    assertThat(entity.getUsername()).isEqualTo("alice");
    assertThat(entity.getName()).isEqualTo("title");
    assertThat(entity.getDescription())
      .isEqualTo("description");
  }
}
```

This code can be described as follows:

- `CoreDomainTest`: This is the name of this test suite. By convention, test suite classes end with the word `Test`. It's not uncommon for them to end with `UnitTest` for unit tests, `IntegrationTest` for integration tests, and any other qualifiers.

- `@Test`: This JUnit annotation signals that this method is a test case. Be sure to use the `org.junit.jupiter.api` version of `@Test`, and not the `org.junit` version. The former package is JUnit 5, while the latter package is JUnit 4 (both are on the classpath to support backward compatibility).

- `newVideoEntityShouldHaveNullId`: The name of the test method is important as it should convey the gist of what it verifies. This isn't a technical requirement but instead an opportunity to capture information. This method verifies that when we create a new instance of `VideoEntity`, its `id` field should be `null`.

- The first line of the method creates an instance of `VideoEntity`.

- `assertThat()`: An AssertJ static helper method that takes a value and verifies it with a collection of clauses.

- `isNull()`: This verifies that the `id` field is `null`.

- `isEqualTo()`: This verifies that the various fields are equal to their expected values.

Inside our IDE, we can right-click the class and run it:

Figure 5.1 – Right-clicking on a test class and running it

Upon running the test suite, you will see the results:

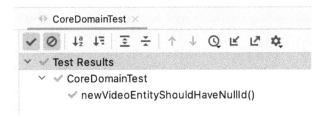

Figure 5.2 – Seeing the test results (green checkmark for passing)

Cut off from this output is the fact that this test case took about 49 milliseconds to run. Running tests with frequency is critical when it comes to adopting a test philosophy. Every time we edit some code, we should run our test suites – if possible, all of them.

Before we move on to more testing techniques, remember how we agreed that any public-facing method should be tested? This extends to things such as the toString() method of the domain class, shown as follows:

```
@Test
void toStringShouldAlsoBeTested() {
  VideoEntity entity = new VideoEntity("alice", "title",
    "description");
  assertThat(entity.toString())
    .isEqualTo("VideoEntity{id=null, username='alice',
      name='title', description='description'}");
}
```

This test method can be described as follows:

- @Test: Again, this annotation is used to indicate this is a test method.
- toStringShouldAlsoBeTested(): Always try to use test method names as a way to capture test intent. Tip: I always like using *should* somewhere in the method name to hone in on its purpose.
- Again, the first line creates an instance of VideoEntity with barebones information.
- assertThat(): This is used to verify whether the value of the toString() method has the expected value.

> **To combine assertions or not to combine assertions?**
>
> This test method's assertion could arguably be added to the previous test method. After all, they both have the same VideoEntity. Why split it out into separate methods? To very clearly capture the intent of testing the entity's toString() method. The previous test method focuses on populating an entity using its constructor and then checking out its getter methods. The toString() method is a separate method. By breaking assertions out into smaller test methods, there's less chance of one failing test masking another.

To round things out, let's verify our domain object's setters:

```
@Test
void settersShouldMutateState() {
  VideoEntity entity = new VideoEntity("alice", "title",
    "description");
  entity.setId(99L);
  entity.setName("new name");
  entity.setDescription("new desc");
  entity.setUsername("bob");
  assertThat(entity.getId()).isEqualTo(99L);
  assertThat(entity.getUsername()).isEqualTo("bob");
  assertThat(entity.getName()).isEqualTo("new name");
  assertThat(entity.getDescription()) //
    .isEqualTo("new desc");
}
```

This code can be described as follows:

- settersShouldMutateState(): This test method is aimed at verifying the entity's setter methods

- The first line creates the same entity instance as the other test cases

- The test method then proceeds to exercise all of the entity's setter methods

- It uses the same AssertJ assertions as before, but with different values, verifying that the state was mutated properly

With this test class in place, we are in a position to engage in test coverage. **IntelliJ** (and most modern IDEs) offer the means to run the test cases with coverage utilities, as shown here:

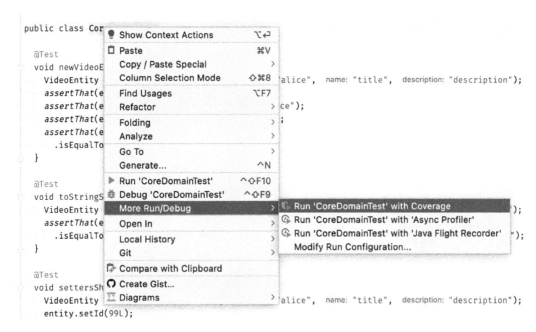

Figure 5.3 – Running test class with coverage analysis

IntelliJ shows which lines have been tested with color highlights. It reveals that our `VideoEntity` entity class is fully covered except for the protected no-argument constructor. It's left as an exercise for you to write another test case verifying that constructor.

This section has shown you how to write unit-level tests against a domain class. It's easy to extend this concept into classes that contain functions, algorithms, and other functional features.

But another area we must take into consideration is the fact that the bulk number of applications out there are web applications. And the next section will show various tactics to verify Spring MVC web controllers.

Testing web controllers with MockMVC

With web pages being a key component of a web app, Spring comes loaded with tools to easily verify web functionality.

While we could instantiate a Spring MVC web controller and interrogate it using various assertions, this would get clunky. And part of what we seek is going through the machinery of Spring MVC. Essentially, we need to make a web call and wait for a controller to respond.

To test the HomeController class we built earlier in this book, we need to create a new test class, HomeControllerTest.java, underneath src/test/java in the same package as HomeController:

```
@WebMvcTest(controllers = HomeController.class)
public class HomeControllerTest {

  @Autowired MockMvc mvc;

  @MockBean VideoService videoService;

  @Test
  @WithMockUser
  void indexPageHasSeveralHtmlForms() throws Exception {
    String html = mvc.perform( //
      get("/")) //
      .andExpect(status().isOk()) //
      .andExpect( //
        content().string( //
          containsString("Username: user"))) //
      .andExpect( //
          content().string( //
          containsString("Authorities: [ROLE_USER]"))) //
      .andReturn() //
      .getResponse().getContentAsString();

    assertThat(html).contains( //
      "<form action=\"/logout\"", //
      "<form action=\"/search\"", //
      "<form action=\"/new-video\"");
  }
}
```

This tiny test class can be explained as follows:

- @WebMvcTest: A Spring Boot test annotation that enables Spring MVC's machinery. The controllers parameter constrains this test suite to only the HomeController class.

- `@Autowired MockMvc mvc`: `@WebMvcTest` adds an instance of Spring's `MockMvc` utility to the application context. Then, we can autowire it into our test suite for all test methods to use.

- `@MockBean VideoService videoService`: This bean is a required component of `HomeController`. Using Spring Boot Test's `@MockBean` annotation creates a mocked version of the bean and adds it into the application context.

- `@Test`: Denotes this method as a JUnit 5 test case.

- `@WithMockUser`: This annotation from **Spring Security Test** simulates a user logging in with a username of `user` and an authority of `ROLE_USER` (default values).

- The first line uses `MockMvc` to perform a `get("/")`.

- The subsequent clauses perform a series of assertions, including verifying whether the result is an `HTTP 200 (OK)` response code and that the content contains a username of `user` and an authority of `ROLE_USER`. Then, it wraps up the MockMVC call by grabbing the entire response as a string.

- Following the MockMVC call is an AssertJ assertion, verifying bits of HTML output.

This test method essentially invokes the base URL of the `HomeController` class and checks various aspects of it, such as the response code as well as its content.

A key feature of our web app is the ability to create new videos. We are engaging with the HTML form we put on the web page earlier in this chapter through the method shown below:

```
@Test
@WithMockUser
void postNewVideoShouldWork() throws Exception {
  mvc.perform( //
    post("/new-video") //
        .param("name", "new video") //
        .param("description", "new desc") //
        .with(csrf())) //
    .andExpect(redirectedUrl("/"));

  verify(videoService).create( //
    new NewVideo( //
      "new video", //
      "new desc"), //
    "user");
}
```

This test method can be described as follows:

- `@Test`: A JUnit 5 annotation to signal this as a test method.

- `@WithMockUser`: Gets us past Spring Security's checks by simulated authentication.

- This time, the test method uses MockMVC to perform `post("/new-video")` with two parameters (`name` and `description`). Since the web page uses **Cross-Site Request Forgery (CSRF)**, we can use `.with(csrf())` to automatically supply the proper CSRF token, further simulating this as a valid request and not an attack.

- `redirectedUrl("/")`: This verifies that the controller issues an HTTP redirect.

- `verify(videoService)`: Mockito's hook to verify that the `create()` method of the mocked `VideoService` bean was called with the same parameters fed by MockMVC and the username from `@WithMockUser`.

With all this created, we can easily run our test suite, as shown here:

✓ Test Results	195 ms
✓ HomeControllerTest	195 ms
✓ postNewVideoShouldWork()	126 ms
✓ indexPageHasSeveralHtmlForms()	69 ms

Figure 5.4 – Test results from HomeControllerTest

This screenshot of test results shows us successfully exercising a couple of the controller's methods in sub-second time.

It's left as an exercise for you to test out the other controller methods.

Being able to prove fundamental controller behavior this quickly is critical. It allows us to build up a regime of tests, verifying *all* our controllers. And as mentioned earlier in this chapter, the more tests we write, the more confidence we can instill in our system.

Something we briefly touched on in this section was using a mocked-out `VideoService` bean. There is a lot more we can do with mocking, as we'll cover in the next section.

Testing data repositories with mocks

Having run our web controller through some automated testing, it's time to switch our attention to another key piece of our system: the **service layer** that the web controller invokes.

Something that's key is spotting any collaborators. Since the only service that's injected into `HomeController` is `VideoService`, let's take a closer look.

`VideoService`, as defined in *Chapter 3*, *Querying for Data with Spring Boot*, has one collaborator, `VideoRepository`. Essentially, to test out the `VideoService` bean in a unit-test fashion, we need to isolate it from any outside influences. This can be accomplished using mocking.

Unit testing versus integration testing

There are various test strategies we can leverage. A key one is unit versus integration testing. In principle, a **unit test** is meant to only test one class. Any external services should be mocked or stubbed out. The counterpart test strategy, **integration testing**, involves creating real or simulated variants of these various collaborators. Naturally, there are benefits and costs associated with both. Unit testing tends to be faster since all outside influences are swapped out with canned answers. But there is the risk that a given test case doesn't test anything but the mock itself. Integration testing can increase confidence, since it tends to be more *real*, but it also tends to take more design and setup. And whether you're using embedded databases or **Docker** containers to emulate production services, the services aren't as fast. This is why any genuine application tends to have a mixture of both. Some amount of unit testing can verify core functionality. But we also need a sense that when our components are connected, they operate correctly together.

In the previous section, we leveraged Spring Boot Test's slice-based `@WebMvcTest` annotation. In this section, we will use a different tactic to configure things:

```
@ExtendWith(MockitoExtension.class)
public class VideoServiceTest {

  VideoService service;
  @Mock VideoRepository repository;

  @BeforeEach
  void setUp() {
    this.service = new VideoService(repository);
  }
}
```

This test class can be described as follows:

- `@ExtendWith(MockitoExtension.class)`: Mockito's JUnit 5 hook to mock out any fields with the `@Mock` annotation
- `VideoService`: The class under test
- `VideoRepository`: A collaborator required for `VideoService` is marked for mocking

- @BeforeEach: JUnit 5's annotation to make this setup method run before every test method

- The setUp() method shows VideoService being created with the mock VideoRepository injected through its constructor

Mockito has always had its static mock() method, which makes it possible to create mock objects. But using their @Mock annotation (and the MockitoExtension JUnit 5 extension) makes it very clear which component is under test.

With this machinery in place, we are ready to add our first test method:

```
@Test
void getVideosShouldReturnAll() {
  // given
  VideoEntity video1 = new VideoEntity("alice", "Spring
    Boot 3 Intro", "Learn the basics!");
  VideoEntity video2 = new VideoEntity("alice", "Spring
    Boot 3 Deep Dive", "Go deep!");
  when(repository.findAll()).thenReturn(List.of(video1,
    video2));

  // when
  List<VideoEntity> videos = service.getVideos();

  // then
  assertThat(videos).containsExactly(video1, video2);
}
```

This test method has some key parts, as follows:

- @Test : Once again, this is JUnit 5's annotation that flags it as a test method.

- The first two lines are about creating some test data. The third line uses Mockito to define how the mock VideoRepository responds when its findAll() method is invoked.

- The next line is where the getVideos() method of VideoService is called.

- The last line uses AssertJ to verify the outcome.

These bullet points, while accurate, don't capture the entire flow of this test method.

For starters, this method has three comments denoting three phases: given, when, and then. The given, when, and then concept is a staple behind **behavior-driven design** (**BDD**). The idea is that given a set of inputs, when you do an action *X*, then you can expect *Y*.

Test cases that flow this way tend to be easier to read. And not just by software developers, but also by business analysts and other teammates who aren't as focused on writing code but instead focused on capturing customer intent.

> **Tip**
> There's no requirement to include comments, but this convention makes it easier to read. And it's not just the comments. Sometimes, we can write test cases that are all over the place. Making a test method follow `given`, `when`, and `then` can help make them more cogent and focused. For example, if a test case seems to have too many assertions and veers off in too many directions, it could be a sign that it should be broken up into multiple test methods.

We would be remiss if we didn't mention that Mockito includes a series of matching operators. See the following test case, where we're testing the ability to create new video entries:

```
@Test
void creatingANewVideoShouldReturnTheSameData() {
  // given
  given(repository.saveAndFlush(any(VideoEntity.class)))
    .willReturn(new VideoEntity("alice", "name", "des"));

  // when
  VideoEntity newVideo = service.create
    (new NewVideo("name", "des"), "alice");

  // then
  assertThat(newVideo.getName()).isEqualTo("name");
  assertThat(newVideo.getDescription()).isEqualTo("des");
  assertThat(newVideo.getUsername()).isEqualTo("alice");
}
```

The key parts to note are as follows:

* `given()`: This test method uses Mockito's `BDDMockito.given` operator, a synonym for Mockito's `when()` operator

* `any(VideoEntity.class)`: Mockito's operator to match when the repository's `saveAndFlush()` operation is called

* The middle of the test method shows us invoking `VideoService.create()`

* The test method wraps up, asserting against the results

Mockito's `BDDMockito` class also has a `then()` operator that we could have used instead of the assertions. This hinges on whether we're testing data or behavior!

Testing data versus testing behavior

A given test case generally consists of either verifying against test data or verifying that the right methods were called. So far, we've used `when(something).thenReturn(value)`, which is known as a **stub**. We are configuring a set of canned test data to be returned for a specific function call. And later, we can expect to assert these values. The alternative is to use Mockito's `verify()` operator, which we'll see in the next test case. This operator, instead of confirming by data, checks what method was called on the mock object. We don't have to commit to one strategy or another. Sometimes, with the code we're testing, it is easier to capture its intent through stubbing. Other times, it's clearer to capture behavior with mocking. Either way, Mockito makes it easy to test.

While `BDDMockito` provides nice alternatives, it's easier (at least to me) to simply use the same operators everywhere. Whether we are stubbing or mocking depends on the test case.

Check out the following final test case, where we are verifying our service's `delete` operation:

```
@Test
void deletingAVideoShouldWork() {
  // given
  VideoEntity entity = new VideoEntity("alice", "name",
    "desc");
  entity.setId(1L);
  when(repository.findById(1L))
    .thenReturn(Optional.of(entity));

  // when
  service.delete(1L);

  // then
  verify(repository).findById(1L);
  verify(repository).delete(entity);
}
```

This test method has some key differences from the previous ones:

- `when()`: Since Mockito's `given()` operator is just a synonym, it's easier to use the same `when()` operator everywhere.

- This test invokes the `delete()` operation of `VideoService`.

- `Verify()`: Because the behavior of the service is more complex, canned data won't work. Instead, we must switch to verifying the methods invoked inside the service.

It should be pointed out that entire books have been written about Mockito. My friend, Ken Kousen, recently wrote *Mockito Made Clear* (see `https://springbootlearning.com/mockito-book`), which I'd recommend for a deeper dive.

We are merely scratching the surface of the sophistication this toolkit affords us. Suffice it to say, we have captured an appreciable amount of the `VideoService` API through readable test scenarios.

However, one thing is key in all this: these test cases are *unit*-based. And that comes with certain limitations. To grow our confidence, we will expand our testing reach through the use of an in-memory database in the next section.

Testing data repositories with embedded databases

Testing against real databases has always been expensive in terms of both time and resources. That's because it has traditionally required launching our application, grabbing a handwritten script of sorts, and clicking through various pages of the application to ensure it works.

There are companies with teams of test engineers whose sole job is to write these test documents, update them as changes are rolled out, and run them against applications in carved-out test labs.

Imagine waiting a week for your new feature to get checked out by this regimen.

Automated testing brought a new wave of empowerment to developers. They can capture test cases describing the scenario they were aiming for. Yet developers still ran into the issue of talking to a *real* database (because let's face it – tests aren't real unless you're talking to a physical database) until people started developing databases that could speak SQL yet run locally and in-memory.

> **Don't all databases run in memory?**
>
> Production-grade database systems run in memory. Servers are specced out with giant memory and disk space amounts to support a database server. But this isn't what we are talking about. An **in-memory database** concerning your application is a database that runs in the same memory space as your application.

There are a handful of choices. For this section, we'll use **HyperSQL Database** (**HSQLDB**). We can choose this from Spring Initializr at `https://start.spring.io` and add it to our build project with the following Maven coordinates:

```
<dependency>
    <groupId>org.hsqldb</groupId>
```

```
        <artifactId>hsqldb</artifactId>
        <scope>runtime</scope>
</dependency>
```

This dependency has one key aspect: it's a `runtime` one, meaning that nothing in our code has to compile against it. It's only needed when the application runs.

Now, to test against `VideoRepository`, which we built back in *Chapter 3*, *Querying for Data with Spring Boot*, create the `VideoRepositoryHsqlTest` class below `src/test/java`, in the related package:

```
@DataJpaTest
public class VideoRepositoryHsqlTest {

  @Autowired VideoRepository repository;

  @BeforeEach
  void setUp() {
    repository.saveAll( //
      List.of( //
        new VideoEntity( //
          "alice", //
          "Need HELP with your SPRING BOOT 3 App?", //
          "SPRING BOOT 3 will only speed things up."),
        new VideoEntity("alice", //
          "Don't do THIS to your own CODE!", //
          "As a pro developer, never ever EVER do this to
            your code."),
        new VideoEntity("bob", //
          "SECRETS to fix BROKEN CODE!", //
          "Discover ways to not only debug your code")));
  }
}
```

This test class can be described as follows:

* `@DataJpaTest`: This is Spring Boot's test annotation and indicates that we want it to perform all of its automated scanning of entity class definitions and Spring Data JPA repositories.

- `@Autowired VideoRepository`: Automatically injects an instance of our `VideoRepository` object to test against.

- `@BeforeEach`: This is JUnit 5's annotation and ensures that this method runs before each test method.

- `repository.saveAll()`: Using `VideoRepository`, it saves a batch of test data.

With this set up, we can start drafting test methods to exercise our various repository methods. Now, it's important to understand that we aren't focused on confirming whether or not Spring Data JPA works. That would imply we are verifying the framework, a task outside our scope.

No – we need to verify that we have written the correct queries, whether that's using custom finders, query by example, or whatever other strategies we wish to leverage.

So, let's write the first test:

```
@Test
void findAllShouldProduceAllVideos() {
  List<VideoEntity> videos = repository.findAll();
  assertThat(videos).hasSize(3);
}
```

This test method does the following:

- Exercises the `findAll()` method.

- Using AssertJ, it checks the size of the results. We could dig a little deeper into the assertions (we'll do this later in this section!).

It's left as an exercise for you to expand this test method (after reading further down) to comprehensively verify the data.

Part of our `search` feature was to do a case-insensitive check for one video. Write a test for that like this:

```
@Test
void findByNameShouldRetrieveOneEntry() {
  List<VideoEntity> videos = repository //
    .findByNameContainsIgnoreCase("SpRinG bOOt 3");
  assertThat(videos).hasSize(1);
  assertThat(videos).extracting(VideoEntity::getName) //
    .containsExactlyInAnyOrder( //
      "Need HELP with your SPRING BOOT 3 App?");
}
```

This test method is more extensive and can be explained as follows:

- Notice how the name of the test method gives us a sense of what it does?
- It uses `findByNameContainsIgnoreCase()` and plugs in a jumbled-up sub-string.
- Using AssertJ, it verifies the size of the results as 1.
- Using AssertJ's `extracting()` operator and a **Java 8** method reference, we can extract the name field of each entry.
- The last portion of this assertion is `containsExactlyInAnyOrder()`. If the order doesn't matter, but the specific contents do, then this is a perfect operator for confirming results.

A question you may have is: why aren't we asserting against `VideoEntity` objects? After all, **Java 17** records make it super simple to instantiate instances of them.

The reason to dodge this in a test case, especially one that talks to a real database, is that the `id` field is populated by the `saveAll()` operation in the `setUp()` method. While we could brainstorm ways to dynamically deal with this between `setUp()` and a given test method, it's not critical that we work to confirm primary keys.

Instead, focus on trying to verify things from the application's perspective. In this situation, we want to know that our mixed-cased partial input yields the correct video and verify that the `name` field fits the bill perfectly.

Another test we can write is confirming that searching by name or description works. So, add the following test method:

```
@Test
void findByNameOrDescriptionShouldFindTwo() {
  List<VideoEntity> videos = repository //
    .findByNameContainsOrDescriptionContainsAllIgnoreCase(
      "CoDe", "YOUR CODE");
  assertThat(videos).hasSize(2);
  assertThat(videos) //
    .extracting(VideoEntity::getDescription) //
    .contains("As a pro developer, never ever EVER do this
      to your code.", //
      "Discover ways to not only debug your code");
}
```

This test method can be described as follows:

- Here, we are exercising the repository's `findByNameContainsOrDescription ContainsAllIgnoreCase()`. The inputs are indeed partial strings, and the case is altered from what was stored in the `setUp()` method.

- Again, asserting the size of the results is an easy test to verify we are on the right path.

- This time, we are using the `extracting()` operator to fetch the `description` field.

- We simply check that the extracting operator contains a couple of descriptions without worrying about the order. It's important to remember that, without an `ORDER BY` clause, databases are not obligated to return the results in the same order as they were stored.

One thing should be pointed out: this test class used **field injection** to autowire `VideoRepository`. In modern Spring apps, it's usually recommended to use **constructor injection**. We saw this in more detail in *Chapter 2*, *Creating a Web Application with Spring Boot*, in the *Injecting dependencies through constructor calls* section.

While field injection is usually seen as a risk that might lead to null pointer exceptions, when it comes to test classes, it's alright. That's because the life cycle for creating and destroying test classes is handled by JUnit and neither us nor Spring Framework.

Now, there is still one repository method we have yet to test, and that's the `delete()` operation. We'll cover that later in this chapter when we explore *Testing security policies with Spring Security Test*.

In the meantime, we must visit a critical issue lingering right before us: what if our target database isn't embedded?

If we were to use something more mainstream in production such as PostgreSQL, MySQL, MariaDB, Oracle, or some other relational database, we have to deal with the fact that they are not available as embedded, co-located processes.

We could continue using HSQL as a basis for writing test cases. And even though we're using JPA as a standard, we still run the risk of our SQL operations not working properly when we get to production.

Even though SQL is a standard (or rather, multiple standards), there are gaps not covered by the specs. And every database engine fills those gaps with its solutions as well as offers features outside the specs.

This leads us to a need to write test cases against, say, PostgreSQL, but not be able to use what we've seen so far. And this leads us to the next section.

Adding Testcontainers to the application

We have seen that, with mocking, we can replace a *real* service with a fake one. But what happens when you need to verify a *real* service, which involves talking to a real database?

The fact that each database engine has slight variations in implementations of SQL demands that we test our database operations against *the same version* we intend to use in production!

With the emergence of **Docker** in 2013 and the rise of putting various tools and applications inside containers, it has become possible to find a container for the database we seek.

Further cultivated by open source, just about every database we can find has a containerized version.

While this makes it possible for us to spin up an instance on our local workstation, the task of manually launching a local database every time we want to run our tests doesn't quite cut it.

Enter **Testcontainers**. With their first release coming out in 2015, Testcontainers provides a mechanism to start up a database container, invoke a series of test cases, and then shut down the container. All with no manual action from you or me.

To add Testcontainers to any Spring Boot application, again, we only need to visit Spring Initializr at `start.spring.io`. From there, we can select **Testcontainers**, as well as **PostgreSQL Driver**.

The changes to add to our `pom.xml` build file are shown here:

```
<testcontainers.version>1.17.6</testcontainers.version>
```

`testcontainers.version` specifies the version of Testcontainers to use. This property setting should be placed inside the `<properties/>` element, the same place where you can find the already existing `java.version` property.

With that in place, the following dependencies must also be added:

```
<dependency>
        <groupId>org.postgresql</groupId>
        <artifactId>postgresql</artifactId>
        <scope>runtime</scope>
</dependency>
<dependency>
        <groupId>org.testcontainers</groupId>
        <artifactId>postgresql</artifactId>
        <scope>test</scope>
</dependency>
<dependency>
        <groupId>org.testcontainers</groupId>
        <artifactId>junit-jupiter</artifactId>
        <scope>test</scope>
</dependency>
```

These additional dependencies can be described as follows:

- `org.postgresql:postgresql`: A third-party library managed by Spring Boot. This is the driver to connect to a PostgreSQL database; hence, it only needs to be `runtime` scoped. There is nothing in our code base that must compile against it.

- `org.testcontainers:postgresql`: The Testcontainers library that brings in first-class support for PostgreSQL containers (which we'll explore further in this section).

- `org.testcontainers:junit-jupiter`: The Testcontainers library that brings deep integration with JUnit 5 – that is, JUnit Jupiter.

It's important to understand that Testcontainers involves a fleet of various modules, all managed under the umbrella of GitHub repositories. They do this by releasing a Maven **Bill of Materials** (**BOM**), a central artifact that contains all the versions.

The `testcontainers.version` property specifies what version of Testcontainers BOM we wish to use, which is added to the `pom.xml` file in a separate section below `<dependencies/>`, as shown here:

```
<dependencyManagement>
    <dependencies>
        <dependency>
            <groupId>org.testcontainers</groupId>
            <artifactId>testcontainers-bom</artifactId>
            <version>${testcontainers.version}</version>
            <type>pom</type>
            <scope>import</scope>
        </dependency>
    </dependencies>
</dependencyManagement>
```

This BOM entry can be described as follows:

- `org.testcontainers:testcontainers-bom`: This Testcontainers BOM contains all the key information about each supported module. By specifying the version here, all other Testcontainers dependencies can skip setting a version.

- `pom`: A dependency type that indicates this artifact has no code, only Maven build information.

- `import`: A scope indicating that this dependency shall be replaced effectively by whatever this BOM contains. It's a shortcut for adding a stack of declared versions.

With all this set up, we can write some test cases in the next section!

Testing data repositories with Testcontainers

The first step when it comes to using Testcontainers is configuring the test case. To talk to a Postgres database, do this:

```
@Testcontainers
@DataJpaTest
@AutoConfigureTestDatabase(replace = Replace.NONE)
public class VideoRepositoryTestcontainersTest {

  @Autowired VideoRepository repository;

  @Container //
  static final PostgreSQLContainer<?> database = //
    new PostgreSQLContainer<>("postgres:9.6.12") //
      .withUsername("postgres");
}
```

This framework for test cases can be described as follows:

- `@Testcontainers`: The annotation from the Testcontainers `junit-jupiter` module that hooks into the life cycle of a JUnit 5 test case.

- `@DataJpaTest`: Spring Boot Test's annotation we used in the previous section, indicating that all entity classes and Spring Data JPA repositories should be scanned.

- `@AutoConfigureTestDatabase`: This Spring Boot Test annotation tells Spring Boot that instead of swapping out the `DataSource` bean like it normally does, to instead NOT replace it like it normally does when there's an embedded database on the classpath (more on why this is needed shortly).

- `@Autowired VideoRepository`: Injects the application's Spring Data repository. We want the real thing and not some mock because this is what we're testing!

- `@Container`: Testcontainer's annotation to flag this as the container to control through the JUnit life cycle.

- `PostgreSQLContainer`: Creates a Postgres instance through Docker. The constructor string specifies the Docker Hub coordinates of the exact image we want. Note that this makes it easy to have multiple test classes, each focused on different versions of Postgres!

All of this allows us to spin up a real instance of Postgres and leverage it from a test class. The extra annotations unite the start and stop actions of Docker with the start and stop actions of our test scenario.

However, we aren't *quite* there.

Spring Boot, with its autoconfiguration magic, either creates a real `DataSource` bean or an embedded one. If it spots H2 or HSQL on the testing classpath, it pivots toward using the embedded database. Otherwise, it has some autoconfiguration default settings based on which JDBC driver it sees.

And neither situation is what we want. We want it to switch away from H2 and HSQL and instead use Postgres. But the hostname and port will be wrong since this isn't standalone Postgres but instead Docker-based Postgres.

Never fear, `ApplicationContextInitializer` comes to the rescue. This is the Spring Framework class that grants us access to the startup life cycle of the application, as shown here:

```
static class DataSourceInitializer //
  implements ApplicationContextInitializer
    <ConfigurableApplicationContext> {
  @Override
  public void initialize(ConfigurableApplicationContext
    applicationContext) {
    TestPropertySourceUtils.
      addInlinedPropertiesToEnvironment(applicationContext,
      "spring.datasource.url=" + database.getJdbcUrl(),
      "spring.datasource.username="+database.getUsername(),
      "spring.datasource.password="+database.getPassword(),
      "spring.jpa.hibernate.ddl-auto=create-drop");
  }
}
```

This code can be explained as follows:

- `ApplicationContextInitializer<ConfigurableApplicationContext>`: This class is what gives us a handle on the application context.

- `initialize()`: This method is the callback Spring will invoke while the application context is getting created.

- `TestPropertySourceUtils.addInlinedPropertiesToEnvironment`: This static method from Spring Test allows us to add additional property settings to the application context. The properties provided here are from the `PostgreSQLContainer` instance created in the previous section. We'll be tapping into a container already started by Testcontainers so that we can harness its JDBC URL, `username`, and `password`.

- `spring.jpa.hibernate.ddl-auto=create-drop`: When talking to an embedded database, Spring Boot autoconfigures things with JPA's `create-drop` policy, where the database schema is created from scratch. Because we are using a *real* connection to talk to a PostgreSQL database, it switches to none, where none of Spring Boot's embedded behavior happens. Instead, Spring Boot will attempt to make no changes to the database regarding schema and data. Since this is a test environment, we need to override this and switch back to `create-drop`.

To apply this set of properties by hooking the Testcontainers-managed database into Spring Boot's autoconfigured `DataSource`, we simply need to add the following to the test class:

```
@ContextConfiguration(initializers = DataSourceInitializer.
  class)
public class VideoRepositoryTestcontainersTest {
     ...
}
```

The `@ContextConfiguration` annotation adds our `DataSourceInitializer` class to the application context. And due to it registering an `ApplicationContextInitializer`, it will be invoked at precisely the right moment after Testcontainers has launched a Postgres container and before Spring Data JPA autoconfiguration is applied.

The only thing left to do is write some actual tests!

Since each test method starts with a clean database, we need to pre-load some content, as shown here:

```
@BeforeEach
void setUp() {
  repository.saveAll( //
    List.of( //
      new VideoEntity( //
        "alice", //
        "Need HELP with your SPRING BOOT 3 App?", //
        "SPRING BOOT 3 will only speed things up."),
      new VideoEntity("alice", //
        "Don't do THIS to your own CODE!", //
        "As a pro developer, never ever EVER do this to
          your code."),
      new VideoEntity("bob", //
        "SECRETS to fix BROKEN CODE!", //
        "Discover ways to not only debug your code")));
}
```

This method can be described as follows:

- `@BeforeEach`: The JUnit annotation that runs this code *before* each test method.
- `repository.saveAll()`: This stores a whole list of `VideoEntity` objects in the database.
- `List.of()`: A Java 17 operator to quickly and easily assemble a list.
- Each `VideoEntity` instance has a `user`, a `name`, and a `description`.

What if we need to test different sets of data? Different data-driven scenarios? Write another test class! You can use Testcontainers easily between different test classes. By tightly integrating with JUnit, there is no need to sweat some static instance floating around from one test class, wrecking this test class.

Now, with all this set up, we can *finally* write some tests, as shown here:

```
@Test
void findAllShouldProduceAllVideos() {
    List<VideoEntity> videos = repository.findAll();
    assertThat(videos).hasSize(3);
}
```

This test method verifies that the `findAll()` method returns all three entities stored in the database. Considering `findAll()` is provided by Spring Data JPA, this is bordering on testing Spring Data JPA and not our code. But sometimes, we need this type of test to simply verify we've set everything up correctly.

This is also sometimes called a **smoke test**, a test case that verifies things are up and operational.

A more in-depth test case involves proving that our custom finder that supports our search feature is working, as shown here:

```
@Test
void findByName() {
    List<VideoEntity> videos = repository.
        findByNameContainsIgnoreCase("SPRING BOOT 3");
    assertThat(videos).hasSize(1);
}
```

This test method has the same annotations and AssertJ annotations, but it focuses on that same `findByNameContainsIgnoreCase`, using data stored in a database.

To wrap things up, let's verify our uber-long custom finder with a test case, as shown here:

```
@Test
void findByNameOrDescription() {
  List<VideoEntity> videos = repository.
    findByNameContainsOrDescriptionContainsAllIgnoreCase
    ("CODE", "your code");
  assertThat(videos).hasSize(2);
}
```

Yikes! That method name is so long it wrecks the formatting in this book. This could be a sign that this scenario is hankering for **Query by Example**. Time to go back to *Chapter 3*, *Querying for Data with Spring Boot*, and consider replacing this query, perhaps?

With all this in place, we can run our test suite and confidently know that our data repository properly interacts with the database. Not only do we know that our repository is doing things right, but our test methods are also meant to ensure our system is doing the right thing.

These tests verify that our design of case-insensitive queries against various fields supports the preceding service layer:

✔ Test Results	460 ms
✔✔ VideoRepositoryTestcontainersTest	460 ms
✔ findAllShouldProduceAllVideos()	401 ms
✔ findByName()	35 ms
✔ findByNameOrDescription()	24 ms

Figure 5.5 – Testcontainers-based tests

And while this section focused on a repository connecting to a database, this tactic works in many other places – RabbitMQ, Apache Kafka, Redis, Hazelcast, anything. If you can find a Docker Hub image, you can hook it into your code via Testcontainers. Sometimes, there are shortcut annotations. Other times, you just need to create the container yourself like we just did.

Having verified our web controller, our service layer, and now our repository layer, there is just one thing to tackle: verifying our security policy.

Testing security policies with Spring Security Test

Has something crossed your mind? Didn't we check out security stuff when we wrote that `HomeControllerTest` class earlier in this chapter?

Yes… and no.

We used the `@WithMockUser` annotation from Spring Security Test earlier in this chapter. But that's because any `@WebMvcTest`-annotated test class will, by default, have our Spring Security policies in effect.

But we didn't cover all the necessary security paths. And in security, there are often many paths to cover. And as we dig, we'll discover exactly what this means.

For starters, we need a new test class, as shown here:

```
@WebMvcTest (controllers = HomeController.class)
public class SecurityBasedTest {

  @Autowired MockMvc mvc;

  @MockBean VideoService videoService;
}
```

Hopefully, things are starting to look familiar:

- `@WebMvcTest`: This Spring Boot Test annotation indicates this is a web-based test class focused on `HomeController`. It's important to understand that Spring Security policies will be in effect.

- `@Autowired MockMvc`: Automatically injects a Spring MockMVC instance for us to craft test cases.

- `@MockBean VideoService`: `HomeController`'s collaborator is to be replaced by a Mockito mock.

With this in place, we can start by verifying access to the home page. In this context, it makes sense to inspect our `SecurityConfig`:

```
http.authorizeHttpRequests() //
  .requestMatchers("/login").permitAll() //
  .requestMatchers("/", "/search").authenticated() //
  .requestMatchers(HttpMethod.GET, "/api/**").
    authenticated()
  .requestMatchers(HttpMethod.POST, "/delete/**",
    "/new-video").authenticated() //
  .anyRequest().denyAll() //
  .and() //
  .formLogin() //
```

```
  .and() //
  .httpBasic();
```

This list of security rules has one bold-faced rule toward the top. It indicates that access to / requires authenticated access and nothing more.

To verify that unauthenticated users are denied access, write the following test case:

```
@Test
void unauthUserShouldNotAccessHomePage() throws Exception {
  mvc //
    .perform(get("/")) //
    .andExpect(status().isUnauthorized());
}
```

This test method has some key aspects:

- It does *NOT* have one of those `@WithMockUser` annotations. This means no authentication credentials are stored in the servlet context, thus simulating an unauthorized user.

- `mvc.perform(get("/"))`: Use MockMVC to perform a GET / call.

- `status().isUnauthorized()`: This asserts that the result is an HTTP 401 Unauthorized error code.

Also, note the method name of the test: unauthUserShouldNotAccessHomePage. It very clearly states the expectation. That way, if it ever breaks, we'll know exactly what the point of the test was. Hopefully, this puts us on the path toward fixing things faster.

> **status().isUnauthorized() for an unauthenticated user?**
>
> In security, proving who you are is called **authentication**. What you are allowed to do is known as **authorization**. However, the HTTP status code for an *unauthenticated* user is **401 Unauthorized**. When someone is authenticated but attempts to access something they aren't *authorized* for, the HTTP status code is **403 Forbidden**. A rather quirky mixture of terminology, but something to be aware of.

We just wrote a *bad* path test case, a critical requirement when testing security policies. We need to also write a *good* path test case, as shown here:

```
@Test
@WithMockUser(username = "alice", roles = "USER")
void authUserShouldAccessHomePage() throws Exception {
  mvc //
```

```
    .perform(get("/")) //
    .andExpect(status().isOk());
}
```

This test method is very similar except for the following:

- @WithMockUser: This annotation inserts an authentication token into the MockMVC servlet context with a username of alice and an authority of ROLE_USER.

- It does the same get("/") call the previous test method did but expects a different outcome. With status().isOk(), we are looking for an HTTP 200 Ok result code.

We have now rounded out our test regime to verify that the home page is properly locked down. However, unauthenticated users and ROLE_USER users aren't the only users our system has. We also have administrators with ROLE_ADMIN. And for each role, we really should have a separate test to ensure our security policy is properly configured.

The following code is almost the same as the preceding code:

```
@Test
@WithMockUser(username = "alice", roles = "ADMIN")
void adminShouldAccessHomePage() throws Exception {
  mvc //
    .perform(get("/")) //
    .andExpect(status().isOk());
}
```

The only difference is that @WithMockUser has alice and ROLE_ADMIN stored in the servlet context.

These three tests should properly verify access to the home page.

Considering our HomeController also affords us the ability to add new video objects, we should also write some test methods to ensure things are handled properly, as shown here:

```
@Test
void newVideoFromUnauthUserShouldFail() throws Exception {
  mvc.perform( //
    post("/new-video") //
      .param("name", "new video") //
      .param("description", "new desc") //
      .with(csrf())) //
      .andExpect(status().isUnauthorized());
}
```

This test method can be described as follows:

- The method name clearly describes that it's to verify that an unauthorized user does *NOT* create a new video. Again, this method has no `@WithMockUser` annotation.

- `mvc.perform(post("/new-video"))`: Uses MockMVC to perform a `POST` `/new-video` action. The `param("key", "value")` arguments let us provide the fields normally entered through an HTML form.

- `with(csrf())`: We have CSRF protections enabled. This additional setting lets us hook in the CSRF value, simulating a *legitimate* access attempt.

- `status().isUnauthorized()`: Ensures that we get an HTTP `401` `Unauthorized` response.

If you supply all the expected values, including a valid CSRF token, it will fail, as expected.

> **CSRF**
>
> In *Chapter 4, Securing an Application with Spring Boot*, we found out that Spring Security automatically enables CSRF token checking on forms and other actions to avoid CSRF attacks. For test cases where we don't have CSRF tokens rendered in HTML pages, we must still present this value to avoid shutting off CSRF.

Now, let's write a test where the user has the right permissions to create a new video:

```
@Test
@WithMockUser(username = "alice", roles = "USER")
void newVideoFromUserShouldWork() throws Exception {
  mvc.perform( //
    post("/new-video") //
      .param("name", "new video") //
      .param("description", "new desc") //
      .with(csrf())) //
      .andExpect(status().is3xxRedirection()) //
      .andExpect(redirectedUrl("/"));
}
```

This code can be summarized as follows:

- `@WithMockUser`: This user has `ROLE_USER`.

- It performs the same `POST` `/new-video` with the same values and CSRF tokens, yet we get a different set of response codes.

- `status().is3xxRedirection()`: Verifies that we get something in the 300 series of HTTP response signals. This makes our test case less brittle if, say, someone switches from soft redirects to hard redirects in the future.

- `redirectedUrl("/")`: Lets us verify that the redirected path is /.

This test method machinery is identical to the previous test method. The only difference is the setup (`alice/ROLE_USER`) and the outcomes (redirect to /).

And this is what makes these test methods security-focused. The point here is to see that accessing the same endpoints but with different credentials (or none at all) yields proper results.

Thanks to MockMVC and Spring Security Test, it's easy to exercise the Spring MVC machinery and assert against it. And thanks to Spring Boot Test, it's super easy to activate actual parts of our application, again building confidence.

Summary

Throughout this chapter, we have explored multiple ways to write test cases. We have seen simple tests, midgrade tests, and complex ones. All of these give us ways to test different aspects of our application.

And each tactic has various tradeoffs. We can get our hands on real database engines if we're willing to spend the extra runtime. We can also ensure our security policies are properly shaken out with both unauthorized and fully authorized users.

Hopefully, this has whet your appetite to fully embrace testing in your applications.

In the next chapter, *Configuring an Application with Spring Boot*, we'll learn how to parameterize, configure, and override settings for our application.

Part 3:
Releasing an Application
with Spring Boot

Building an application is only half the battle. Releasing the application to production is critical. You will learn how to configure your application for various environments, including the cloud. You will also discover different ways to package it up and get it into your customers' hands. Finally, you will learn how to ramp things up with native images.

This part includes the following chapters:

- *Chapter 6, Configuring an Application with Spring Boot*
- *Chapter 7, Releasing an Application with Spring Boot*
- *Chapter 8, Going Native with Spring Boot*

6

Configuring an Application with Spring Boot

In the previous chapter, we learned how to test various aspects of an application, including web controllers, repositories, and domain objects. We also explored security-path testing, as well as using Testcontainers to emulate production.

In this chapter, we'll learn how to configure our application, which is a critical piece of application development. While at first glance this may sound like setting a handful of properties, there is a deeper concept at play.

Our code needs a connection to the real world. In this sense, we're talking about anything our application connects to: databases, message brokers, authentication systems, external services, and more. The details needed to point our application at a given database or message broker are contained in these property settings. By making application configuration a first-class citizen in Spring Boot, application deployment becomes versatile.

The point of this chapter is to reveal how application configuration becomes not just simple, but instead, a tool to make applications better serve our needs. This way, we can spend all our time serving the application's needs!

In this chapter, we'll cover the following topics:

- Creating custom properties
- Creating profile-based property files
- Switching to YAML
- Setting properties with environment variables
- Order of property overrides

> **Where to find this chapter's code**
>
> The source code for this chapter can be found at `https://github.com/PacktPublishing/Learning-Spring-Boot-3.0/tree/main/ch6`.

Creating custom properties

We have already dabbled with application properties in a couple of places in this book. Remember setting `spring.mustache.servlet.expose-request-attributes=true` in the `application.properties` file in *Chapter 4, Securing an Application with Spring Boot*?

Configuring our application using property files is incredibly handy. While Spring Boot offers many custom properties we can use, it's possible to create our own!

Let's start by creating some custom properties. To do that, create a brand new `AppConfig` class, like this:

```
@ConfigurationProperties("app.config")
public record AppConfig(String header, String intro,
  List<UserAccount> users) {
}
```

This Java 17 record can be described as follows:

- `@ConfigurationProperties`: A Spring Boot annotation that flags this record as a source of property settings. The `app.config` value is the prefix for its properties.

- `AppConfig`: The name of this bundle of type-safe configuration properties. It doesn't matter what name we give it.

  The fields themselves are the names of the properties, which we'll cover in more detail shortly.

This tiny record essentially declares three properties: `app.config.header`, `app.config.intro`, and `app.config.users`. We can populate them right away by adding the following to `application.properties`:

```
app.config.header=Greetings Learning Spring Boot 3.0 fans!
app.config.intro=In this chapter, we are learning how to make a
web app using Spring Boot 3.0
app.config.users[0].username=alice
app.config.users[0].password=password
app.config.users[0].authorities[0]=ROLE_USER
app.config.users[1].username=bob
app.config.users[1].password=password
```

```
app.config.users[1].authorities[0]=ROLE_USER
app.config.users[2].username=admin
app.config.users[2].password=password
app.config.users[2].authorities[0]=ROLE_ADMIN
```

This batch of properties can be described as follows:

- `app.config.header`: A string value to insert into the top-most part of our template (which we'll do soon).

- `app.config.intro`: A string greeting to put in the template.

- `app.config.users`: A list of `UserAccount` entries with each attribute split out into a separate line. The square bracket notation is used to populate the Java list.

These property settings are cool, but we don't have access to them *yet*. We need to enable them by adding a little more. In truth, it doesn't matter where so long as it's on a Spring component we know will be picked up by Spring Boot's component scanning.

Since this set of properties (you can have more than one!) is application-wide, why not apply them to the entry point for our application?

```
@SpringBootApplication
@EnableConfigurationProperties(AppConfig.class)
public class Chapter6Application {
  public static void main(String[] args) {
    SpringApplication.run(Chapter6Application.class, args);
  }
}
```

This code is very similar to the code in previous chapters with one change:

- `@EnableConfigurationProperties(AppConfig.class)`: This annotation activates this application configuration, making it possible to inject into any Spring bean.

The rest of the code is the same as we've seen in previous chapters.

> **Tip**
>
> Where is the best place to enable custom properties? In truth, it doesn't matter. So long as it gets enabled on some Spring bean that will be picked up by Spring Boot, it will get added to the application context. However, if this particular set of properties were specific to a certain Spring bean, it's recommended to put the annotation on that bean definition, emphasizing that the bean comes with a complement of configurable properties. If the properties are used by more than one bean, consider what we just did.

Thanks to @EnableConfigurationProperties, a Spring bean of the AppConfig type will be registered automatically in the application context, bound to the values applied inside application. properties.

To leverage it in our HomeController, we only need to make the following changes at the top of the class:

```
@Controller
public class HomeController {

    private final VideoService videoService;
    private final AppConfig appConfig;

    public HomeController(VideoService videoService,
        AppConfig appConfig) {
            this.videoService = videoService;
            this.appConfig = appConfig;
    }
        ...rest of the class...
```

HomeController has one change: a field of the AppConfig type is declared below VideoService and it gets initialized in the constructor call.

With this in place, we can use its provided values in rendering the index template further down in HomeController, like this:

```
@GetMapping
public String index(Model model, //
    Authentication authentication) {
    model.addAttribute("videos", videoService.getVideos());
    model.addAttribute("authentication", authentication);
    model.addAttribute("header", appConfig.header());
    model.addAttribute("intro", appConfig.intro());
    return "index";
}
```

These changes can be described as follows:

- The model's "header" attribute is populated with appConfig.header()
- The model's "intro" attribute is populated with appConfig.intro()

This will take the string values we put into `application.properties` and route them so that they render `index.mustache`.

To complete the loop, we need to make the following changes to the template:

```
<h1>{{header}}</h1>
<p>{{intro}}</p>
```

We simply use Mustache's double-curly syntax to grab the `{{header}}` and `{{intro}}` attributes. What used to be hard-coded is now a template variable!

This is neat, but is parameterizing a couple of fixed values in a template that big a deal? Perhaps not. To get our hands on the power of Spring Boot configuration properties, let's delve into the `users` field.

That one is *NOT* ready to go. Why?

It's important to understand the Java property fields are fundamentally built up out of key-value pairs of strings. The values aren't wrapped in double quotes, but they are pretty much treated that way.

Spring comes with some convenient converters built in, but centered inside the `AppConfig` user's property, inside the `UserAccount` type, is a `List<GrantedAuthority>`. Converting strings into `GrantedAuthority` is not clear-cut and requires that we write and register a converter.

The code for handling user accounts is security-focused, so it makes sense to register this custom converter inside our already-existing `SecurityConfig`:

```
@Bean
@ConfigurationPropertiesBinding
Converter<String, GrantedAuthority> converter() {
    return new Converter<String, GrantedAuthority>() {
        @Override
        public GrantedAuthority convert(String source) {
            return new SimpleGrantedAuthority(source);
        }
    };
}
```

This code can be described as follows:

- `@Bean`: This converter must be registered with the application context so that it gets picked up properly.

- `@ConfigurationPropertiesBinding`: Due to application property conversion happening very early in the application life cycle, Spring Boot *ONLY* applies converters that have this annotation applied. Try to avoid pulling in other dependencies.

- `Convert ()`: At the heart is one little method that converts a `String` into a `GrantedAuthority` by using `SimpleGrantedAuthority`.

Spring converters are quite handy. However, you may be encouraged by your IDE to *simplify* this code a little bit. It may suggest that you convert this into a **Java lambda expression** (`return source -> new SimpleGrantedAuthority(source)`), or perhaps trim it down into a **method reference** (`return SimpleGrantedAuthority::new`).

But that won't work. At least, not without some help.

That's because Java has **type erasure**. At runtime, Java will drop the generic information, which makes it impossible for Spring to find the proper converter and apply it. So, we must either keep things just the way they are or adopt a different strategy, as shown here:

```
interface GrantedAuthorityCnv extends Converter<String,
GrantedAuthority> {}
@Bean
@ConfigurationPropertiesBinding
GrantedAuthorityCnv converter() {
  return SimpleGrantedAuthority::new;
}
```

This type-erasure beating solution can be described as follows:

- `GrantedAuthorityCnv`: By extending Spring's `Converter` interface with a custom interface that applies our generic parameters, we are freezing a copy of these parameters that Spring can find and use

- Using this new interface instead of `Converter<String, GrantedAuthority>`, we can switch to the slimmed-down method reference

This may appear to be roughly the same amount of code. It is simply a different way the same information is being put together. It's a personal preference whether a fully-expanded `Converter<String, GrantedAuthority>` inside a bean definition or a slimmed-down but separate interface is easier to grok.

Either way, we can now fire up our application, knowing that we have varying aspects of our application retooled to be property-driven.

And with all this in place, we can begin to explore how a property-driven application gives us the ability to customize things for different environments in the next section.

Creating profile-based property files

In the previous section, we realized the ability to extract certain aspects of our application into property files. The next big step you can take is realizing how far you can go with this.

Inevitably, we run into situations such as carrying our application into a new environment and wondering, "can we change the properties for *THIS* situation?"

For example, what if our application, before release, has to be installed in a test bed where it can get checked out? The databases are different. The test team may want a different set of test accounts. And any other external services (message brokers, authentication systems, and so on) will probably also be different.

So, the question arises, "can I have a different set of properties?" To which Spring Boot says "yes!"

To check this out, create another property file. Call it `application-test.properties` and load it up, like this:

```
app.config.header=Greetings Test Team!
app.config.intro=If you run into issues while testing, let me
know!
app.config.users[0].username=test1
app.config.users[0].password=password
app.config.users[0].authorities[0]=ROLE_NOTHING
app.config.users[1].username=test2
app.config.users[1].password=password
app.config.users[1].authorities[0]=ROLE_USER
app.config.users[2].username=test3
app.config.users[2].password=password
app.config.users[2].authorities[0]=ROLE_ADMIN
```

These alternative sets of properties can be described as follows:

- `application-test.properties`: When you append `-test` to the base name of the property file, you can activate it by using the Spring `test` profile
- The web bits are tuned to the audience
- The test team would probably have a set of users they want to carry out all their scenarios

To run the application, all we have to do is activate the Spring `test` profile. There are multiple ways to switch that one:

- Add `-Dspring.profiles.active=test` to the JVM

- In a Unix environment, use export SPRING_PROFILES_ACTIVE=test

- Some IDEs even support this directly! Check out the following screenshot from IntelliJ IDEA (Ultimate Edition, not Community Edition):

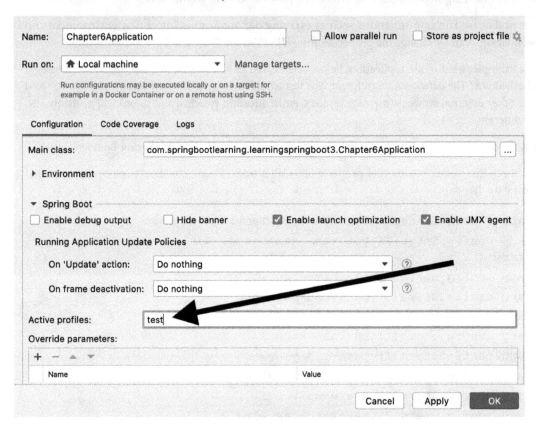

Figure 6.1 – IntelliJ's Run dialog

Look toward the bottom for **Active profiles** and see where we entered test.

Want to check out IntelliJ IDEA Ultimate Edition?

JetBrains does offer a 30-day free trial for IntelliJ IDEA Ultimate Edition. They also have options if you happen to be working on certain open source projects. Personally, this is my favorite IDE I've used throughout my career. It has incredible insight into all kinds of things from Java code to SQL strings in Spring Data @Query annotations to property files to Spring Boot powerups (and more). Feel free to check out your options at https://springbootlearning. com/intellij-idea-try-it.

When we activate a Spring profile, Spring Boot will add `application-test.properties` to our configuration.

> **Tip**
>
> Profiles are additive. When we activate the `test` profile, `application-test.properties` is added to the configuration. It doesn't replace `application.properties`. However, if a given property is found in both, the last profile applied wins. Considering `application.properties` is considered the default profile, `application-test.properties` would be applied afterward. Hence, the `header`, `intro`, and `users` properties would be overrides. It's also possible to apply more than one profile, separated by commas.

Imagine if we had A) a development lab where we had a slimmed-down complement of servers, B) a full-sized test bed with a separate set of servers, and C) a production environment with full-sized servers. We could consider the following:

* Make production the *default* configuration and put connection URLs pointed at our production servers in `application.properties`. Also, have a counterpart `test` profile captured in the `application-test.properties` file with connection URLs for the test servers. Finally, make individual developers use a `dev` profile, using the `application-dev.properties` file with connection URLs for the development lab of servers.

* Another tactic is to flip production and development. When anyone runs the application by default, with no custom profiles activated, it runs in `dev` mode. To run things in production, activate the `production` profile and apply `application-production.properties`. Keep the `test` profile the same as the previous example.

So far, we've talked about different environments and real, physical environments. Of course, we could be discussing traditional servers in various rack configurations. But we could just as well be talking about virtualized servers. Either way, adjusting connection URLs among other settings through the use of environment-specific configuration files is valuable.

But that's not the only choice put before us.

What if our application is being deployed in the cloud? In different clouds? What if we started with a traditional set of hardware, as discussed earlier in this section, but management has decided to migrate to AWS, Azure, or VMware Tanzu?

There is no need to go into our code and start making changes. Instead, all we need to do is work on a new property file and plug in whatever the connection details are to communicate with our cloud-based services!

Switching to YAML

Part of the *Spring way* is options. Developers have varying needs based on circumstances, and Spring tries to offer different ways to effectively get things done.

And sometimes, the number of property settings we need can explode. With the property file's key/value paradigm, this can get unwieldy. In the previous section, where we had lists and complex values, it became clunky to have to specify index values.

YAML is a more succinct way to represent the same settings. Perhaps an example is in order. Create an `application-alternate.yaml` file in the `src/main/resources` folder, as shown here:

```
app:
  config:
    header: Greetings from YAML-based settings!
    intro: Check out this page hosted from YAML
    users:
      -
        username: yaml1
        password: password
        authorities:
          - ROLE_USER
      -
        username: yaml2
        password: password
        authorities:
          - ROLE_USER
      -
        username: yaml3
        password: password
        authorities:
        - ROLE_ADMIN
```

These YAML-based settings can be described as follows:

- The nested nature of YAML prevents duplicate entries and makes it clear where each property is located

- The hyphens underneath `users` denote array entries

- Because the `users` field of `AppConfig` is a complex type (`List<UserAccount>`), each field of each entry is listed on a separate line

- Because `authorities` itself is a list, it too uses hyphens

In *Chapter 4*, *Securing an Application with Spring Boot*, in the *Leveraging Google to authenticate users* section, we had a first-hand look at using YAML to configure Spring Security OAuth2 settings.

YAML is slim and trim because it avoids duplicating entries. In addition to that, its nested nature makes things a little more readable.

> **Tradeoff**
>
> Everything has tradeoffs, right? I've seen many systems built on top of YAML configuration files, especially in the cloud configuration space. If your YAML file is so big that it doesn't fit on one screen, the nice readable format can work against you. That's because the spacing and tabbing between nested levels are important. That example I just gave was about something not specific to Spring Boot. The number of properties you'll be dealing with will probably be fine. Just don't be surprised if you work on something else and discover the dark side of YAML.

A bonus feature provided by most modern IDEs is code completion support for both `application.properties` files and `application.yaml` files!

Check out the following screenshot:

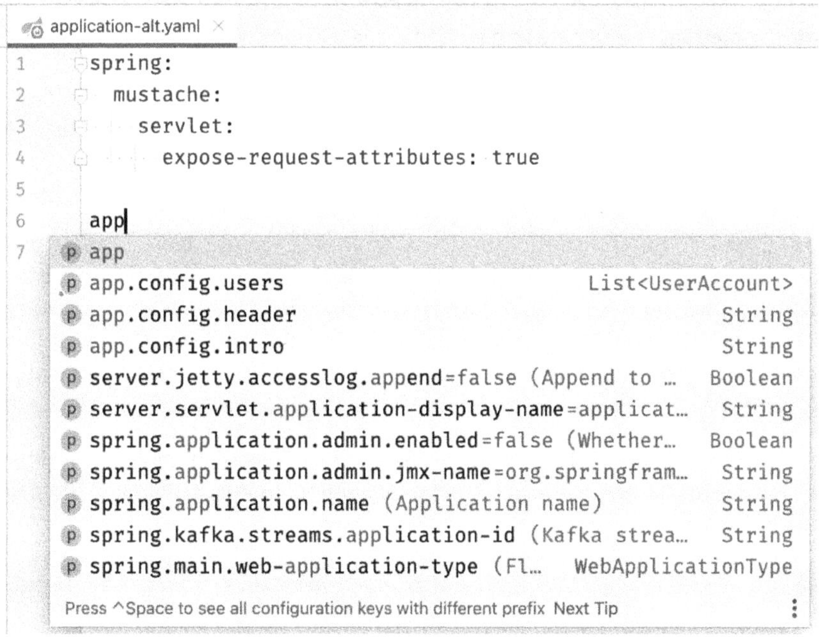

Figure 6.2 – IntelliJ's code completion for property settings

In this screenshot, while the available properties appear with standard key/value notation, they will be automatically applied in YAML format if need be.

Spring Boot encourages us to use custom, type-safe configuration property classes. To support having our configuration settings appear in code completion popups like the one shown previously, all we must do is add the following dependency to our pom.xml file!

```xml
<dependency>
        <groupId>org.springframework.boot</groupId>
        <artifactId>spring-boot-configuration-processor
          </artifactId>
        <optional>true</optional>
</dependency>
```

But this isn't the only way to supply property settings. In the next section, we'll learn how environment variables can also be used to override property settings.

Setting properties with environment variables

A configurable application wouldn't be complete if there weren't a way to configure it straight from the command line. This is of key value because no matter how much thought and design we put into our applications, something will always pop up.

Being stuck with a bundled-up application and no way to override the various property files stuffed inside it would be a showstopper.

> **Don't do this!**
>
> Maybe you've run into situations where you need to unpack the JAR file, edit some property files, and bundle it back up. Do *not* do this! This is a hack that may have skated by 20 years ago, but it just doesn't cut it today. In today's age of controlled pipelines and secured release processes, it's simply too risky to manually get your hands on a JAR file and tweak it like that. And thanks to the real-world experience of the Spring team, there's no *need* to do that.

You can easily override any property using an environment variable. Remember how, earlier in this chapter, we activated a profile using IntelliJ IDEA's Run dialog? You can do the same thing right from the command line, like this:

```
$ SPRING_PROFILES_ACTIVE=alternative ./mvnw spring-boot:run
```

This command can be explained as follows:

- `SPRING_PROFILES_ACTIVE`: The alternative way to reference properties on a Mac/Linux-based system. Dots typically don't work as well, so capitalized tokens with underscores work.

- `alternative`: The profile we are running. In fact, in the console output, you can see **The following 1 profile is active: "alternative"** as proof that it's running with that profile activated.

- `./mvnw`: Run things using the Maven wrapper. This is a handy way to use Maven without having to install it on your system (really handy for CI systems!).

- `spring-boot:run`: The Maven command that activates the `run` goal of `spring-boot-maven-plugin`.

It's also important to understand that the environment variables, when used this way, only apply to the current command. To make environment variables persistent for the duration of the current shell, you have to export the variable (which we won't cover here).

And it's easy to activate multiple profiles. Just separate the profile names with commas, as shown here:

```
$ SPRING_PROFILES_ACTIVE=test,alternate ./mvnw spring-boot:run
```

This is almost the same as activating a single profile, except both the `test` and `alternate` profiles have been activated.

Considering both the `test` and the `alternate` profiles defined a different set of users, perhaps you're wondering which ones are active. It's simple – the property files are applied from left to right.

Since the `alternate` profile goes last, it will layer on any new properties while overriding any duplicates. Thus, the YAML-based accounts are what end up getting configured!

That isn't the only rule when it comes to property overrides and options. Just read the next section to grok all the options we have to tune properties.

Ordering property overrides

Back in *Chapter 1*, *Core Features of Spring Boot*, we got a summary of the order of property settings.

Let's see that list of options again:

- Default properties are provided by Spring Boot's `SpringApplication.setDefaultProperties()` method.

- `@PropertySource`-annotated `@Configuration` classes.

- Config data (such as `application.properties` files).

- A `RandomValuePropertySource` that has properties only in `random.*`.

- OS environment variables.

- Java System properties (`System.getProperties()`).

- JNDI attributes from `java:comp/env`.

- `ServletContext` init parameters.

- `ServletConfig` init parameters.

- Properties from `SPRING_APPLICATION_JSON` (inline JSON embedded in an environment variable or system property).

- Command-line arguments, as shown in the previous section.

- `properties` attribute on your tests. Available with the `@SpringBootTest` annotation and also slice-based testing (which we covered back in *Chapter 5*, *Testing with Spring Boot* in the *Testing data repositories with Testcontainers* section)

- `@TestPropertySource` annotations on your tests.

- DevTools global settings properties (the `$HOME/.config/spring-boot` directory when Spring Boot DevTools is active).

This list is ordered from lowest priority to highest priority. The `application.properties` file is pretty low, meaning it offers a great way to set a baseline for our properties, but we have multiple ways to override them either in tests or at deployment. Further down the list are all the ways we can override that baseline.

In addition to that, configuration files are considered in the following order:

- Application properties packaged inside your JAR file

- Profile-specific application properties inside your JAR file

- Application profiles outside your JAR file

- Profile-specific application properties outside your JAR file

That's right. We can have application property files sitting adjacent to our runnable JAR file and have them serve as overrides. Remember that earlier warning to not crack open a JAR and tweak its properties?

No need! You can simply create a new property file and apply any adjustments.

Warning

There is still some risk with adjusting properties on the fly from the command line. Anytime you do, take note and consider capturing these changes in your version control system, perhaps as a different profile. Nothing is worse than hammering out a solution and watching it get overwritten with a patch that didn't pick up your changes.

There is a fundamental concept in software engineering that encircles this idea of decoupling code from configuration through the power of property settings. It's known as a **Twelve-Factor App**, a concept that arose back in 2011 from cloud vendor *Heroku* (now owned by *Salesforce*). To be specific, *configuration* is the *third factor* listed at `https://12factor.net/`.

The idea is to externalize everything likely to vary from environment to environment. It extends the life of our application when we design with this sort of flexibility. Our examples of property settings may have been a bit contrived, but hopefully, you can see the value in externalizing certain bits.

And with Spring Boot's ability to pull together profile-specific properties, it becomes even easier to apply this feature to our code.

Whether all the factors of Twelve-Factor Apps are relevant today or applicable to our next application is debatable. But many of those factors listed on the site certainly lend themselves to making our applications easier to deploy, link, and stack up, like building blocks to build systems. So, I'd consider reading through it in your spare time.

Summary

In this chapter, we learned how to externalize parts of our system from content that appears in the web layer to the list of users allowed to authenticate with the system. We saw how to create type-safe configuration classes, bootstrap them from property files, and then inject them into parts of our application. We even saw how to use profile-based properties and choose between using traditional Java property files or using YAML. We then explored even more ways to override property settings from the command line and checked out a comprehensive list of yet more ways to override properties.

While our examples were probably not as realistic as they could have been, the concept is there. Externalizing properties that are likely to vary from environment to environment is a valuable feature, and Spring Boot eases the usage of this pattern.

In the next chapter, *Releasing an Application with Spring Boot*, we are finally going to release our application into the wild. There, we'll learn what Spring Boot does to not only help us perform this task but then proceed to manage our application on Day 2.

7

Releasing an Application with Spring Boot

In the previous chapter, we learned about all the various ways we could configure our application using Spring Boot. This unlocked the ability to run our application in multiple environments, which made it more flexible.

The most critical place for our application to be is in production. Otherwise, it's not doing what we set out to do. Production can be a scary place. The Spring team, in its battle-hardened experience, has built many features into Spring Boot to ease the processes involved with assembling applications, staging, and ultimately managing them once deployed.

Building upon the tools covered in the previous chapter, we'll see how Spring Boot can turn what used to be a scary place to be into a thriving environment.

In this chapter, we'll cover the following topics:

- Creating an uber JAR
- Baking a Docker container
- Releasing your application to Docker Hub
- Tweaking things in production

> **Where to find this chapter's code**
>
> This chapter doesn't have much in terms of original code. Instead, the code from *Chapter 6, Configuring an Application with Spring Boot*, has been copied into the following repository so that we can tackle various forms of deployment: `https://github.com/PacktPublishing/Learning-Spring-Boot-3.0/tree/main/ch7`.

Creating an uber JAR

This may or may not sound familiar, but once upon a time, long ago, developers would compile their code, run scripts to assemble the binary bits into ZIP files, and drag them into applications that would ultimately result in burning a CD or staging the file on some retro artifact such as a tape drive or jumbo hard drive.

Then, they would lug that artifact to another location, be it a vaulted room with special access control or an entirely different facility on the other end of town.

This sounds like something out of a post-techno sci-fi movie, to be honest.

But the truth is, the world of production has always been set apart from the world of development, whether we're talking about cubicle farms with dozens of coders on one end of the building and the target server room on the other side of the room, or if we're describing a start-up with five people spread around the world, deploying to Amazon's cloud-based solution.

Either way, the place where our application must live for customers to access it and the place where we develop are two different locations.

And the most important thing is to minimize all the steps it takes to move the code out of our IDE and onto the servers, fielding web requests from around the globe.

That's why, from early on in 2014, the Spring Boot team developed a novel idea: building an **uber JAR**.

The following Maven command is all it takes:

```
% ./mvnw clean package
```

This Maven command has two parts:

- `clean`: Deletes that `target` folder and any other generated outputs. This is always good to include before building an uber JAR to ensure all generated outputs are up-to-date.
- `package`: Invokes Maven's *package* phase, which will cause the verify, compile, and test phases to be invoked in the proper order.

> **Using Windows?**
> The mvnw script only works on Mac or Linux machines. If you're on Windows, you either must have a comparable shell environment, or you can use `./mvnw.cmd` instead. Either way, when you use `start.spring.io` to build your project, you'll have both at hand.

When we use **Spring Initializr** (`https://start.spring.io`), as we did in several of the previous chapters, one of the entries included in our `pom.xml` file was `spring-boot-maven-plugin`, as shown here:

```
<plugin>
        <groupId>org.springframework.boot</groupId>
        <artifactId>spring-boot-maven-plugin</artifactId>
</plugin>
```

This plugin hooks into Maven's *package* phase and does a few extra steps:

1. It grabs the JAR file originally generated by standard Maven packaging procedures (`target/ch7-0.0.1-SNAPSHOT.jar` in this case) and extracts all of its content.

2. Then, it renames the original JAR to set it aside (`target/ch7-0.0.1-SNAPSHOT.jar.original` in this case).

3. It fashions a new JAR file with the original name.

4. In the new JAR file, it adds the Spring Boot loader coder, which is glue code that can read JAR files from within, allowing it to become a runnable JAR file.

5. It adds our application code to the new JAR file in a subfolder called `BOOT-INF`.

6. It adds *ALL* of our application's third-party dependency JARs into this JAR file in a subfolder called `BOOT-INF/lib`.

7. Finally, it adds some metadata about the layers of the application to this JAR file underneath `BOOT-INF` as `classpath.idx` and `layers.idx` (more about that in the next section!).

With nothing but the JVM, we can launch our application, as follows:

```
% java -jar target/ch7-0.0.1-SNAPSHOT.jar

  .   ____          _            __ _ _
 /\\ / ___'_ __ _ _(_)_ __  __ _ \ \ \ \
( ( )\___ | '_ | '_| | '_ \/ _` | \ \ \ \
 \\/  ___)| |_)| | | | | || (_| |  ) ) ) )
  '  |____| .__|_| |_|_| |_\__, | / / / /
 =========|_|==============|___/=/_/_/_/
 :: Spring Boot ::                (v3.0.0)
...etc...
```

With such a simple command, our application has been transformed from an application that runs inside our developer-centric IDE into an application that runs on any machine with the right **Java Development Kit** (JDK).

Maybe this isn't as impressive as you were led to believe, perhaps because this is incredibly simple. So, let's revisit what exactly has happened:

- There is no need to download and install an Apache Tomcat servlet container anywhere. We are using embedded Apache Tomcat. This means this tiny JAR file is carrying the means to run itself.

- There is no need to go through the legacy procedure of installing an application server, fashioning a WAR file, combining it with third-party dependencies using some ugly assembly file to fashion an EAR file, and then uploading the whole thing to some atrocious UI.

- We can push this whole thing to our favorite cloud provider and command the system to run 10,000 copies.

Having Maven output a runnable application only dependent on Java is epic when it comes to deployment.

> **Newsflash**
>
> In the olden days (1997), it was common practice for me to perform a release by walking around (yes, physically walking) to various department heads and having them sign a piece of paper. With that in hand, I would burn a CD with compiled binaries. From there, I would carry out the rest of a 17-page procedure to essentially construct the binaries and walk them into a lab or onto the customer's site and install the software. This process usually took a couple of days. Removing the technical barriers for doing releases like this is groundbreaking.

It should be pointed out that uber JARs weren't invented by the Spring Boot team. The Maven Shade plugin has been around since 2007. This plugin's job is to kind of do the same steps of bundling everything together one JAR file, but differently.

This plugin unpacks all the incoming JAR files, be it our application code or a third-party dependency. All the unpacked files are kind of mixed into a new JAR file. This new file is called a *shaded* JAR.

Some other tools and plugins do the same thing, but they are fundamentally wrong for mixing things this way.

Some applications need to be inside a JAR to work correctly. There is also a risk of the non-class files not ending up in the right place. Apps that consume third-party classes from JAR files may exhibit some aberrant behavior. And there is also a chance you could be violating some library's license as well.

If you go to any library maintainer and ask if they will handle bug reports when you use their released JAR files beyond the scope of the way they are released, you may not get the support expected.

Spring Boot simply lets our code run as if the third-party JAR files are there as always. No shading is required.

But one thing remains: the part where I mentioned that the app is ready to run wherever you have a JDK installed.

What if your target machine doesn't have one?

Check out the next section!

Baking a Docker container

One of the fastest technologies to sweep the tech world has been **Docker**. If you haven't heard of it, Docker is sort of like a virtualized machine *but more lightweight*.

Docker is built on the paradigm of shipping containers. Shipping containers, which are responsible for moving the bulk of the world's goods on ships and trains, have a common shape. This means people can plan how to ship their products while knowing the containers handled by the entire world are all the same size and structure.

Docker is built on top of Linux's libcontainer library, a toolkit that grants not a completely virtual environment, but instead a partially virtual one. It allows a container's processes, memory, and network stack to be isolated from the host server.

Essentially, you install the Docker engine on all your target machines. From there, you are free to install any container, as needed, to do what you need to do.

Instead of wasting time crafting an entire virtual machine, you simply spin up containers on the fly. With Docker's more application-oriented nature, it becomes a much more nimble choice.

And Spring Boot comes with Docker support baked in.

> **Isn't Docker experimental?**
>
> Docker has been around since 2013. When I wrote *Learning Spring Boot 2.0 Second Edition* back in 2017, it was highly experimental at the time. So, I didn't mention it. People were using it on occasion, perhaps for demos or other one-off tasks, but it wasn't widely used for systems in production. Fast-forward to today, and you can see Docker being used *EVERYWHERE*. Every cloud-based provider supports Docker containers. The most popular CI systems let you run container-based jobs. There are a plethora of companies built around the paradigm of orchestrating systems of Docker containers. Atomic Jar, a company built around the Docker-based test platform, Testcontainers, is out there. Asking developers and sys admins to install Docker on both local machines and customer-facing servers is no longer the big ask it once was.

Assuming you have installed Docker (visit `docker.com` to get started) on your machine, this is all it takes!

```
% ./mvnw spring-boot:build-image
```

This time, instead of connecting to any particular phase of Maven's build and deploy life cycle, the Spring Boot Maven plugin will execute a custom task to *bake a container*.

Baking a container is a Docker phrase meaning assembling all the parts needed to run a container. Baking implies that we only need to generate the image of this container once, and we can then reuse it as many times as needed to run as many instances as required.

When Spring Boot does a build-image process, it first runs Maven's package phase. This includes running the standard complement of unit tests. After that, it will assemble the uber JAR we talked about in the previous section.

With a runnable uber JAR in our hands, Spring Boot can then go into the next phase: leveraging **Paketo Buildpacks** to pull the right type of container together.

Building the "right" type of container

What do we mean by the "right" container?

Docker has a caching solution built into it that involves layers. If a given step in building a container sees no changes from the previous container assembly process, it will use the Docker engine's caching layer.

However, if some aspect has changed, it will invalidate the cached layer and create a new one.

Cached layers can include everything from the base image a container is built on (for example, if you were extending a bare Ubuntu-based container) to the Ubuntu packages downloaded to enhance it along with our custom code.

One thing we do *NOT* want to do is mingle our custom application's code with the third-party dependencies our application uses. For example, if we were using Spring Framework 6.0.0's GA release, that is certainly something it would be useful to cache. That way, we don't have to keep pulling it down!

But if our custom code were mingled into the same layer and a single Java file were to change, the whole layer would be invalidated and we'd have to re-pull all of it.

So, putting things such as Spring Boot, Spring Framework, Mustache, and other libraries into one layer, while putting our custom code into a separate layer, is a good design choice.

In the past, this required several manual steps. But the Spring Boot team has made this layered approach the default configuration! Check out the following excerpt of running `./mvnw spring-boot:build-image`:

```
[INFO] --- spring-boot-maven-plugin:3.0.0:build-image (default-
cli) @ ch7 ---
[INFO] Building image 'docker.io/library/ch7:0.0.1-SNAPSHOT'
[INFO]
[INFO]    > Pulled builder image 'paketobuildpacks/builder@
sha256:9fb2c87caff867c9a49f04bf2ceb24c87bde504f3fed88227e9ab5d9
a572060c'
```

```
[INFO]     > Pulling run image 'docker.io/paketobuildpacks/
run:base-cnb' 100%
[INFO]     > Pulled run image 'paketobuildpacks/run@sha256:
fed727f0622994807560102d6a2d37116ed2e03dddef5445119eb0172
12bbfd7'
[INFO]     > Executing lifecycle version v0.14.2
[INFO]     > Using build cache volume 'pack-cache-564d5464b59a.
build'
...
[INFO] Successfully built image 'docker.io/library/ch7:0.0.1-
SNAPSHOT'
```

This small fragment of the total output to the console reveals the following:

- It's using Docker to build an image named `docker.io/library/ch7:0.0.1-SNAPSHOT`. This includes the name of our module as well as the version, both found in our `pom.xml` file.

- It uses the Paketo Buildpack from Docker Hub, as shown by pulling the `paketobuildpacks/builder` and `paketobuildpacks/run` containers.

- It finishes with a successfully assembled container.

Paketo Buildpacks

Paketo Buildpacks (`https://paketo.io/`) is a project focused on turning application source code into container images. Instead of doing it directly, Spring Boot delegates containerization to Paketo. Essentially, it downloads containers that do all the leg work, easing the process for us to bake a container.

Want to see more details?

The complete output of the Spring Boot Maven plugin's build-image task is frankly too long and wide to fit into a book. But you are free to check out the complete output at `https://springbootlearning.com/build-image-output`.

At this stage, we have a fully assembled container. Let's check it out! Using Docker, we can now run the container, as shown here:

```
% docker run -p 8080:8080 docker.io/library/ch7:0.0.1-SNAPSHOT
Calculating JVM memory based on 7163580K available memory
For more information on this calculation, see https://paketo.
io/docs/reference/java-reference/#memory-calculator
```

```
    .    ____                                              _       __ _ _
   /\\ / ___'_ __ _ _(_)_ __  __ _   \ \ \ \
  ( ( )\___ | '_ | '_| | '_ \/ _` | \ \ \ \
   \\/  ___)| |_)| | | | | | || (_| |   ) ) ) )
    '  |____| .__|_| |_|_| |_\__, | / / / /
   =========|_|==============|___/=/_/_/_/
   :: Spring Boot ::                                (v3.0.0)


2022-11-18T23:32:30.711Z       INFO
1 --- [                        main]
c.s.1.Chapter7Application                                        :
Starting Chapter7Application v0.0.1-SNAPSHOT using Java 17.0.5
on 5e4fb7fdead2 with PID 1 (/workspace/BOOT-INF/classes started
by cnb in /workspace)
```

This portion of the console's output can be described as follows:

- `docker run`: The command to run a Docker container

- `-p 8080:8080`: An argument that maps the container's inner port number `8080` to everyone outside the container on port `8080`

- `docker.io/library/ch7:0.0.1-SNAPSHOT`: The name of the container's image

The rest of the output is Docker's output of the running container. The container is now running. In fact, from another shell, we can see it in flight by doing this:

```
% docker ps
CONTAINER ID        IMAGE                                         COMMAND
                        CREATED                    STATUS
         PORTS                                            NAMES
5e4fb7fdead2        ch7:0.0.1-SNAPSHOT        "/cnb/process/web"
5 minutes ago        Up 5 minutes        0.0.0.0:8080->8080/tcp
angry_cray
```

`docker ps` is a command that *shows* any running Docker processes. The output is a bit cramped here inside a book, but this one-line output shows the following:

- `5e4fb7fdead2`: The hashed ID of the container.

- `ch7:0.0.1-SNAPSHOT`: The name of the container's image (without the `docker.io` prefix)

- `/cnb/process/web`: The command the Paketo Buildpack is using to run our Spring Boot application.

- `"5 minutes ago"` and `"Up 5 minutes"`: When the container started and how long it's been up.

- `0.0.0.0:8080->8080/tcp`: The internal to external network mapping.

- `angry_cray`: The *human-friendly* name Docker has given this container's instance. You can refer to an instance by either the hash code or this.

From here, we can shut it down:

```
% docker stop angry_cray
angry_cray
% docker ps
CONTAINER ID     IMAGE          COMMAND      CREATED     STATUS
        PORTS          NAMES
```

This shuts the whole container down. We can see it on the console where we launched everything.

> **Important**
>
> In this instance of spinning up a Docker container, Docker chose the random human-friendly name `angry_cray`. Every container you spin up will have a different, unique, human-friendly name. You can use those details, the container's hashed value, or simply point and click from the Docker Desktop application to control containers on your machine.

All that said, we can now check out the most important step: releasing the container.

Releasing your application to Docker Hub

Building a container is one thing. Releasing that container to production is critical. And as Spring advocate Josh Long likes to say, "Production is the happiest place on Earth!"

You can push the container to your favorite cloud provider. Just about all of them support Docker. But you can also push the container to Docker Hub.

> **Do you have access to Docker Hub?**
>
> Docker Hub offers several plans. You can even get a free account. Your company or university may also grant you access. Be sure to check out `https://docker.com/pricing`, choose the plan that's best for you, and create your account. Assuming you've done so, please check out the rest of this section!

From the console, we can log in directly to our Docker Hub account:

```
% docker login -u <your_id>
Password: *********
```

Assuming all this has been done, we can now push our container to Docker Hub by running the following commands:

```
% docker tag ch7:0.0.1-SNAPSHOT    <user_id>/learning-spring-
boot-3rd-edition-ch7:0.0.1-SNAPSHOT
% docker push <your_id>/learning-spring-boot-3rd-edition-
ch7:0.0.1-SNAPSHOT
```

These steps to push are buried in the following succinct commands. The first involves tagging the local container, as shown here:

- `docker tag <image> <tag>`: Tags the local container name of ch7 possessing the *local* tag of `0.0.1-SNAPSHOT` with a prefix of our Docker Hub *user ID*. This is critical because all of our container images *MUST* match our Docker Hub ID!
- The tagged container also has a name, `learning-spring-boot-3rd-edition-ch7`.
- The tagged container itself has a tag, `0.0.1-SNAPSHOT`.

What just happened?

This can be kind of confusing. Docker Hub containers have three characteristics:

- Name of the container
- Tag of the container
- Namespace of the container

These go together as `namespace/name:tag`. This naming convention does *NOT* have to match whatever your local naming is for the container. You can simply reuse it, but since this is going to be public, you may wish to choose something else.

Tagging is essentially the way to take your local container and give it a *public* name. Whereas there are other Docker repositories, we are sticking with Docker Hub for now as our container repository of choice. And to comply with Docker Hub policy, its `namespace` needs to match our Docker Hub account ID.

What about latest tags?

A common convention seen all over Docker Hub is to use a tag name of latest. This implies that grabbing a container with such a tag would give you, well, the latest release. Or at least the latest *stable* release. But it's important to understand that this is simply a convention. Tags are dynamic and can be moved. So, 0.0.1-SNAPSHOT can also be a dynamic tag you push every time you update the snapshot versions of your application, just like latest. The adoption of Docker Hub by software publishers managing multiple releases has led to multiple tags being managed to indicate which version is being fetched. Before adopting *ANY* container's tag, be sure to check out their tagging strategy so that you can understand what, exactly, you are getting.

Once the container has been tagged, it can be pushed to Docker Hub using the following command:

- `docker push <tagged image>`: Pushes the container to Docker Hub using the public-facing tagged image name

To reiterate the last bullet point, you push the container to Docker Hub using the public-facing name. In our case, we must use `gturnquist848/learning-spring-boot-3rd-edition-ch7:0.0.1-SNAPSHOT`.

Having pushed the container, we can see it on Docker Hub, as shown here:

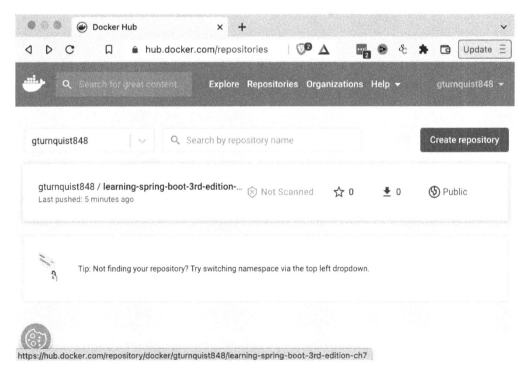

Figure 7.1 – Docker container pushed to Docker Hub

> **Where's my Docker container on Docker Hub?**
>
> The preceding screenshot is from my Docker Hub repository. Any container you push should be on your *own* repository. It's important to use your own Docker Hub ID when tagging and pushing!

We could go much deeper into Docker, Docker Hub, and the world of containers, but frankly, there are entire books dedicated to this topic.

The point of Spring Boot is to make it super simple to wrap up a completed application inside a container and release it to our users, and we just did that *without writing any custom code*.

What would it be like to pick up that container on the other side… and start tweaking things in production? Check out the following section.

Tweaking things in production

An application isn't really in production until we need to start tweaking it, fiddling with it, and making adjustments after release.

This is the very nature of operations. And the various members of the Spring team are no strangers to the world of production.

There are multiple things we can tune and adjust after being handed either an uber JAR or a container. Assuming we have an uber JAR built out of this chapter's code, we can easily type something like this:

```
% java -jar target/ch7-0.0.1-SNAPSHOT.jar
```

This would launch the app with all its default settings, including the standard servlet port of 8080.

But what if we needed it to run next to another Spring Boot web application we just installed yesterday? That suggests we'd need it to listen on a different port. Say no more. All we need do is run a slightly different command, like this:

```
% SERVER_PORT=9000 java -jar target/ch7-0.0.1-SNAPSHOT.jar
...
2022-11-20T15:36:55.748-05:00     INFO 90544 ---
[                    main] o.s.b.w.embedded.tomcat.
TomcatWebServer     : Tomcat started on port(s): 9000 (http)
with context path ''
```

Toward the bottom of the console output, we can see that Apache Tomcat is now listening on port 9000 instead.

That's nice, but it's a bit of a hassle to have to type that extra parameter *every single time*, right?

A better way to offer customized configuration settings is to fashion an additional `application.properties` file *in our local folder*.

First, create a new `application.properties` file, like this:

```
server.port=9000
```

This property override file contains one property: Spring Boot's `server.port` setting with a value of `9000`.

Now, we can run the uber JAR just like we did the first time:

```
% java -jar target/ch7-0.0.1-SNAPSHOT.jar

...
2022-11-20T15:41:09.239-05:00     INFO 91085 ---
[                         main] o.s.b.w.embedded.tomcat.
TomcatWebServer      : Tomcat started on port(s): 9000 (http)
with context path ''
```

This time, when Spring Boot starts up, it looks around and spots the `application.properties` file in our local folder. Then, it applies all its settings as overrides to the one inside the JAR file, and voila!

We have a web app running on port `9000`.

But that is not all. Any property we need to override is up for grabs. We can have multiple override files.

What's an example of this? In the world of constantly evolving requirements, it's not hard to imagine our manager showing up and telling us we need to run not one but *three* instances.

Scaling with Spring Boot

Now, we need to host our application on ports `9000`, `9001`, and `9002` to match up with the load balancer the sysadmins just set up!

Let's expand things and come up with a tactical name for each instance. Something super original, such as `instance1`, `instance2`, and `instance3`.

First, rename that local `application.properties` file to `application-instance1.properties`.

Next, make a copy of the file and name the new one `application-instance2.properties`. Edit the file so that `server.port` is assigned `9001`.

Then, make yet another copy, this time as `application-instance3.properties`. This time, make its `server.port` have a value of `9002`.

With these in place, we can now run three instances using Spring Boot's **profile support**. We'll start by launching instance1, as follows:

```
% SPRING_PROFILES_ACTIVE=instance1 java -jar target/ch7-0.0.1-
SNAPSHOT.jar
...
2022-11-20T15:52:30.195-05:00      INFO 94504 ---
[                        main] o.s.b.w.embedded.tomcat.
TomcatWebServer    : Tomcat started on port(s): 9000 (http)
with context path ''
```

Here, we can see that instance1 is now running on port 9000.

Open another console tab and launch instance2, like this:

```
% SPRING_PROFILES_ACTIVE=instance2 java -jar target/ch7-0.0.1-
SNAPSHOT.jar
...
2022-11-20T15:53:36.403-05:00      INFO 94734 ---
[                        main] o.s.b.w.embedded.tomcat.
TomcatWebServer    : Tomcat started on port(s): 9001 (http)
with context path ''
```

In this console output, we can see instance2 is running on port 9001.

Don't stop there! Let's open a third console tab and run instance3, as shown here:

```
% SPRING_PROFILES_ACTIVE=instance3 java -jar target/ch7-0.0.1-
SNAPSHOT.jar
...
2022-11-20T15:55:53.062-05:00      INFO 96783 ---
[                        main] o.s.b.w.embedded.tomcat.
TomcatWebServer    : Tomcat started on port(s): 9002 (http)
with context path ''
```

Ta-dah!

We now have three instances of our application running on different ports. Buried in the first console's output, we can see the following:

```
2022-11-20T15:52:28.076-05:00      INFO
94504 --- [                        main]
c.s.l.Chapter7Application                               : The
following 1 profile is active: "instance1"
```

This line shows that Spring Boot has spotted that the `instance1` profile is `active`.

There is a similar entry for the other two console outputs. We won't show them here. Suffice it to say, profiles are a powerful way to run multiple instances of what started as a simple, single application.

However, we aren't done. That's because the default configuration for this application happened to be with an in-memory HSQL database. This means that the three instances aren't sharing a common database.

Considering the code was already integration tested against PostgreSQL with Testcontainers, we could tune the settings to allow us to point to a production instance of just such a database!

First, we need to spin up that database. And Testcontainers has shown us the way by using Docker. To run a standalone instance, try this:

```
% docker run -d -p 5432:5432 --name my-postgres -e POSTGRES_
PASSWORD=mysecretpassword postgres:9.6.12
```

This command will spin up a copy of PostgreSQL with the following characteristics:

- `-d`: The instance will run as a background daemon process.

- `-p 5432:5432`: The standard `5432` port will be exported to the public with the same port.

- `--name my-postgres`: The container will run with a fixed name instead of a random one. This will prevent us from running multiple copies at the same time.

- `-e POSTGRES_PASSWORD=mysecretpassword`: The container will run with an environment variable that, according to Postgres's notes, will configure the password.

- `postgres:9.6.12`: The same container coordinates found in the Testcontainers-based integration test.

With this up and running, we can update `application-instance1.properties` with the following additional properties:

```
# JDBC DataSource settings
spring.datasource.url=jdbc:postgresql://localhost:5432/postgres
spring.datasource.username=postgres
spring.datasource.password=mysecretpassword
# JPA settings
spring.jpa.hibernate.ddl-auto=update
spring.jpa.hibernate.show-sql=true
spring.jpa.properties.hibernate.dialect = org.hibernate.
dialect.PostgreSQLDialect
```

The JDBC properties can be summarized as follows:

- `spring.datasource.url`: This is the JDBC connection URL to reach the container-based instance
- `spring.datasource.username`: This contains the default `postgres` username the container runs under
- `spring.datasource.password`: This contains the password we picked earlier in this section

These are all the properties needed for Spring Boot to assemble a JDBC `DataSource` bean.

The JPA properties can be described as follows:

- `spring.jpa.hibernate.ddl-auto`: This is the Spring Data JPA setting that will update the schema if necessary, but not drop or delete anything
- `spring.jpa.hibernate.show-sql`: This will switch on Spring Data JPA's ability to print out SQL statements that get generated
- `spring.jpa.properties.hibernate.dialect`: This is the Hibernate property that signals we are talking to a PostgreSQL-based database

With all these settings, we are aligning JDBC and JPA to communicate with the PostgreSQL database container we just spun up minutes ago.

> **Production data warning!**
>
> The one thing we need to handle… is ensuring that all three instances don't create the same pre-loaded data. In *Chapter 3*, *Querying for Data with Spring Boot*, and *Chapter 4*, *Securing an Application with Spring Boot*, we added some bits to our application that would pre-load user login data, along with some video entries. This is, in fact, best done with outside tools. We should let the DBAs handle setting the schema and loading the data. An application that starts and stops, and runs multiple instances, is *NOT* the vehicle to apply persistent data management policy. Thus, any such beans need to be commented out or at least flagged to *ONLY* run with some *OTHER* profile, such as `setup`. If you examine the final code provided for this chapter, as shown at the beginning of this chapter, you'll find an `application-setup.properties` file, along with such restrictions in the code. We won't show them here, but if you want the app to pre-load this data, run it with profile `setup` (and only do this *ONCE* after starting up the database!).

From here, the sky is the limit. We could run a dozen copies, though we may *NOT* want to do it this way. That's where it becomes valuable to use something meant to orchestrate multiple applications.

There are several options out there, including **Kubernetes** and **Spinnaker**. Kubernetes is a Docker container orchestrator. It lets us manage the containers as well as the load balancers from top to bottom. Check out `https://springbootlearning.com/kubernetes` for more details.

Spinnaker is a continuous delivery pipeline. It makes it possible to take commits to our GitHub repository, package them up into uber JARs, bake Docker container images, and then manage them with rolling upgrades in production. Check out `https://springbootlearning.com/spinnaker` for more information on that.

And of course, there is VMware Tanzu. Tanzu is a complete package, not just a Docker orchestrator. It has solid support for Kubernetes, along with other things. Be sure to check it out at `https://springbootlearning.com/tanzu`.

All of these are powerful tools, each with its tradeoffs. And they provide a comprehensive way to manage Spring Boot applications in production.

Summary

In this chapter, we learned several key skills, including creating an uber JAR that is runnable anywhere, baking a Docker container image that can be run locally with no need for Java, pushing our Docker container to Docker Hub where it can be consumed by our clients, and running multiple instances of our uber JAR pointed at a persistent database, different than the one it came bundled with.

That concludes this chapter! In the next chapter, we will dig in and discover ways we can speed up our Spring Boot application to near-warp speed through the power of GraalVM and something called native applications.

8

Going Native with Spring Boot

In the previous chapter, we learned multiple ways to turn our application from a collection of code into an executable, ready for any production environment, including the cloud. We also learned how to tune it and tweak it so that we could scale it up as needed.

Building upon the tools covered in the previous chapters, we'll see how Spring Boot applications are indeed ready for the future by taking them to some of the most bleeding-edge platforms where performance can be truly dialed up to 11 as we explore native applications.

In this chapter, we'll cover the following topics:

- What is GraalVM and why do we care?

- Retrofitting our application for GraalVM

- Running our native Spring Boot application inside GraalVM

- Baking a Docker container with GraalVM

> **Where to find this chapter's code**
>
> The code for this chapter can be found at `https://github.com/PacktPublishing/Learning-Spring-Boot-3.0/tree/main/ch8`.

The focus of this chapter isn't so much on writing Spring Boot applications as it is about compiling them into a faster, more efficient format (which we'll soon see). Hence, there is no need to write new code. If you check the preceding link, you'll find that this chapter's code is a copy of the previous chapter's code. However, the build file is a little different, which we'll introduce in the next section.

What is GraalVM and why do we care?

For years, Java has suffered a lot of criticism. One of its biggest slights from the early days was its performance. While true to a degree, Java has made quantum leaps by adopting tactics that have allowed it to compete with other platforms at a raw performance level.

Nevertheless, people had continued to criticize Java over things that may seem inane, such as startup time. Indeed, a Java app, running inside its own virtual machine, isn't as fast as Go or C++ binaries. But for the longest time, this has not been an issue, given web apps often have long uptimes.

However, new frontiers of production have exposed this weakness. Continuous deployment systems where 10,000 instances are running all at once and get replaced multiple times a day have made a 30-second cost begin to add up on people's cloud bills.

A new player in the space of production systems has been executable functions. That's right. It's now possible to deploy a single function as an entire application on platforms such as AWS Lambda. And their results could be piped right into another deployed function.

In these scenarios, where the functions are spun up immediately upon demand, things such as startup time *HEAVILY* drive what technologies people use.

And thus has emerged **GraalVM**.

GraalVM by Oracle is essentially a *new virtual machine* that sports support for just about any programming language out there. Don't run your Java JAR files on the JVM. Run them on GraalVM!

GraalVM is a high-performance runtime aimed at Java, JavaScript, Python, Ruby, R, C, and C++. When you're running thousands of instances of systems, then the whole performance output of your applications can make a significant difference.

And the Spring team, in their continuing quest to reduce Java complexity, is here to help. Beginning in 2019, the experimental project Spring Native was born. And since then, almost every facet of the Spring portfolio has been tuned and adjusted to support this endeavor to bring the power of GraalVM to any Spring Boot application.

All with minimal fuss to the end user.

And so, through the rest of this chapter, we will explore taking the application we have come to know in previous chapters and adapting it to the rigors of GraalVM.

Retrofitting our application for GraalVM

There are always two ways to approach building a native application: create a brand-new application or take an existing one and update it. Thanks to Spring Boot 3.0 and their adoption of native application support, it's very easy to update an existing application to use GraalVM instead of the JVM.

> **What is Java Virtual Machine code?**
>
> Java code has always, since the dawn of time, been compiled into **bytecode**, meant to be run on **Java Virtual Machine (JVM)**. This has resulted in the common expression *write once, run anywhere*. Any compiled Java bytecode, due to every aspect of these files being captured by the Java specification, can be run on any compliant JVM, no matter what machine it lives on. This was a huge departure from a previous era that involved compiling separately for every single machine architecture an app would get deployed to. This was revolutionary in its day and has allowed other post-compilation enhancements such as **just-in-time (JIT)** compiler speedups and dynamically making applications slimmer and trimmer.

Compiling applications for GraalVM involves trading in some of that preserved flexibility for faster, more memory-efficient code.

This may trigger a question: why not compile *EVERY ONE* of our applications for GraalVM?

Because of tradeoffs.

GraalVM, to do some of the things it does, requires us to let go of some key features:

- Limited support for reflection
- Limited support for dynamic proxies
- Special handling of external resources

Why? Because GraalVM performs advanced analysis of our code. It uses a concept called *reachability*, where it essentially starts the app and then analyzes what code GraalVM can see. Anything that is *NOT* reachable is simply cut out of the final native image.

Reflection is still possible in native applications. But because not everything is visible directly, it can require extra configuration so that nothing is missed.

Proxies are of a similar issue. Any proxies that are to be supported must be generated at the time of native image building.

This means that accessing bits of code through reflection tactics, deserialization of data, and proxies are trickier and not as straightforward as they once were. The risk is that certain parts of our app may be cut out if we don't properly capture them.

This is one reason why every Spring portfolio project has been diligently working to ensure that whatever bits *NEED* to be in our applications have the necessary *hints* for GraalVM to find them.

> **Beware of faulty information**
>
> Some articles capturing the details of Spring Boot 3.0 and its support for native images may mention that reflection and proxies are simply not supported. This is false. There is support for reflection, but it requires that the code on the other end of such reflective calls is properly registered. Regarding proxies, native images can't handle generating and interpreting bytecode at runtime. All dynamic proxies must be generated at native image build time. These are limitations on the usage of reflection and proxies but not a complete lack of support.

It's also a reason that Spring Framework has slimmed down its usage of reflection tactics to manage the application context. And it's also the reason that Spring Boot has adopted a general approach of not proxying configuration classes containing bean definitions to reduce the number of actual proxies in an application.

For the past 2 years, the Spring team has worked tirelessly with the GraalVM team to tweak and adjust various aspects of the Spring portfolio, removing unnecessary reflection calls and reducing the need for proxies. On top of that, many improvements have been made to GraalVM so that it works better with Spring code.

For us to pick up and run with GraalVM, we will go back to our favorite friend, **Spring Initializr**, at `https://start.spring.io`.

From here, let's start with a fresh set of coordinates:

- **Project**: **Maven**
- **Group**: `com.springbootlearning.learningspringboot3`
- **Artifact**: `ch8`
- **Name**: `Chapter 8`
- **Description**: `Going Native with Spring Boot`
- **Package name**: `com.springbootlearning.learningspringboot3`
- **Packaging**: **Jar**
- **Java**: **17**
- Dependencies:

  - **Spring Web**
  - **Mustache**
  - **H2 Database**
  - **Spring Data JPA**
  - **Spring Security**
  - **GraalVM Native Support**

From here, we can click **EXPLORE**. A popup showing us the build file lets us see what is needed to build a Spring Boot native application.

The Spring Boot starters found include the following:

- `spring-boot-starter-data-jpa`

- `spring-boot-starter-web`

- `spring-boot-starter-mustache`

- `spring-boot-starter-security`

- `h2`

- `spring-boot-starter-test`

Because we're using Spring Data JPA, which involves (by default) Hibernate and its proxied entities, we have this additional plugin:

```
<plugin>
    <groupId>org.hibernate.orm.tooling</groupId>
    <artifactId>hibernate-enhance-maven-plugin</artifactId>
    <version>${hibernate.version}</version>
    <executions>
        <execution>
            <id>enhance</id>
            <goals>
                <goal>enhance</goal>
            </goals>
            <configuration>
                <enableLazyInitialization>
                    true
                        </enableLazyInitialization>
                <enableDirtyTracking>
                    true
                        </enableDirtyTracking>
                <enableAssociationManagement>
                    true
                        </enableAssociationManagement>
            </configuration>
        </execution>
```

```
        </executions>
    </plugin>
```

This helps add some extra settings the Hibernate team identified as critical to Hibernate's proxies working properly with GraalVM.

Something also provided by `spring-boot-starter-parent` (referenced at the top of our build file) is a `native` Maven profile. When we enable it, it changes the settings for `spring-boot-maven-plugin`. Other tools are also brought online, including the **ahead-of-time** (AOT) compilation toolset, as well as GraalVM's `native-maven-plugin`.

We'll see how to utilize all this to build lightning-fast native applications in the next section.

GraalVM and Spring Boot

We're entering a new area of code development. Not only are we talking about building Spring Boot apps, but we're also talking about building them with alternative tools such as GraalVM. You may wish to read Spring Boot's section on *GraalVM Native Image Support* at `https://springbootlearning.com/graalvm`.

Running our native Spring Boot application inside GraalVM

The common convention when building an application for Spring Boot is to run `./mvnw clean package`. This cleans out the old cruft and creates a new uber JAR, something we already saw in *Chapter 7, Releasing an Application with Spring Boot*.

Building a Maven-based project with Spring Boot 3 requires that we have Java 17 installed. But to build a native image, we need to change course.

`native-maven-plugin` mentioned in the previous section, which comes with the `native` Maven profile, requires that we install a different JVM. There are additional tools required to build native images. The easiest way to manage different JVMs on our machine is by using **sdkman** (`https://sdkman.io`).

---

sdkman?

sdkman is an open source tool that allows you to install multiple JDKs and switch between them with ease. It's as easy as `sdk install java 17.0.3-tem` followed by `sdk use java 17.0.3-tem` to download, install, and switch to the Eclipse Foundation's Temurin Java 17.0.3 release. (The Eclipse Foundation is the current maintainer of Jakarta EE.) sdkman is also able to install the right version of the JDK – for example, if you are on an M1 Mac or an older Intel Mac. And in our case, it allows us to install GraalVM's own JDK, which includes all the tools needed to build native images on our machine.

To build native applications on GraalVM, we need to install a version of Java 17 that includes GraalVM tools by typing the following command:

```
% sdk install java 22.3.r17-grl
```

Once it's installed, we can then switch over to it by typing the following:

```
% sdk use java 22.3.r17-grl
```

We can even take a peek at what this version of Java has:

```
% java -version
openjdk version "17.0.5" 2022-10-18
OpenJDK Runtime Environment GraalVM CE 22.3.0 (build
17.0.5+8-jvmci-22.3-b08)
OpenJDK 64-Bit Server VM GraalVM CE 22.3.0 (build
17.0.5+8-jvmci-22.3-b08, mixed mode, sharing)
```

This is **OpenJDK** version 17, also known as Java 17, but it has **GraalVM Community Edition** (CE) version 22.3.0. Essentially, it has all the bits for Java 17 stirred together with GraalVM 22.3.

What is OpenJDK?

OpenJDK is the source of all distributions of Java. SUN Microsystems, the inventors of Java, and later Oracle, initially made the official releases of Java. However, ever since Java 7, all releases of Java start with OpenJDK. Every vendor is free to take the OpenJDK baseline and apply additions as they see fit. However, *ALL* distributions of Java are required to pass a **Technology Compatibility Kit** (TCK) released by Java's executive committee to earn the coffee cup logo. Various vendors offer different levels of support for differing periods and what patches they'll maintain or carry back to the source. There are even distributions from certain vendors that are *NOT* certified against the TCK, so be sure to read all the details before choosing a JDK.

With GraalVM CE's Java 17 active, we can finally build our application natively. To do so, we need to execute the following command:

```
% ./mvnw -Pnative clean native:compile
```

> **Trying to build a native image on Windows?**
>
> Linux is probably the most straightforward platform on which to build native images. Macs have strong support as well, with some gaps in the M1 chipset. To build native images on Windows, though, you need to check out the Windows section of the Spring Boot reference documentation at `https://springbootlearning.com/graalvm-windows`. There, you'll find details about what needs to be installed on your machine to build native images on Windows. Also, don't forget to use `mvnw.cmd` when building with Maven on Windows!

This command will compile our application with the native profile switched on. It leverages `native-maven-plugin` mentioned in the previous section. This process can take a bit longer than building using a standard configuration. And there are a lot of warnings.

The process involves completely scanning the code and performing what's known as AOT compilation. Basically, instead of leaving things in a bytecode format, to be converted into local machine code when the JVM starts up, it instead converts things in advance.

This requires certain features to be curtailed, such as the usage of proxies and reflection. Part of Spring's support in making itself GraalVM-ready was to reduce the usage of proxies and to avoid reflection when not necessary. There are ways to still use such features, but they can bloat up the native executable and remove some of the benefits. The AOT tools also can't see everything on the other side of reflection calls and proxy usage, so they require additional metadata to be registered.

Part of the output can be seen here:

```
[2/7] Performing analysis...  [********]                                    (50.5s @ 5.54GB)
  32,496 (94.29%) of 34,465 classes reachable
  55,941 (71.35%) of 78,401 fields reachable
 156,744 (62.52%) of 250,720 methods reachable
   1,416 classes, 1,822 fields, and 10,190 methods registered for reflection
      67 classes,     75 fields, and     57 methods registered for JNI access
       5 native libraries: -framework CoreServices, -framework Foundation, dl, pthread, z
[3/7] Building universe...                                                   (6.6s @ 4.37GB)
[4/7] Parsing methods...       [**]                                          (3.3s @ 4.85GB)
[5/7] Inlining methods...      [***]                                         (2.0s @ 3.33GB)
[6/7] Compiling methods...     [****]                                       (20.6s @ 5.15GB)
[7/7] Creating image...                                                      (7.3s @ 4.08GB)
  67.76MB (51.26%) for code area:    104,582 compilation units
  63.14MB (47.76%) for image heap:   634,397 objects and 526 resources
   1.29MB ( 0.98%) for other data
 132.20MB in total

Top 10 packages in code area:                  Top 10 object types in image heap:
   2.29MB jdk.proxy4                              14.76MB byte[] for code metadata
   1.66MB sun.security.ssl                         7.93MB java.lang.Class
   1.16MB java.util                                7.35MB byte[] for embedded resources
 914.47KB java.lang.invoke                          6.13MB java.lang.String
 717.60KB com.sun.crypto.provider                   5.63MB byte[] for java.lang.String
 695.62KB org.hibernate.dialect                     4.49MB byte[] for general heap data
 640.51KB org.h2.command                            2.73MB com.oracle.svm.core.hub.DynamicHubCompanion
 638.91KB org.apache.tomcat.util.net                2.02MB byte[] for reflection metadata
 540.62KB org.h2.table                              1.17MB java.lang.String[]
 537.87KB org.apache.catalina.core                  1.11MB c.o.svm.core.hub.DynamicHub$ReflectionMetadata
  57.47MB for 1248 more packages                    8.81MB for 5562 more object types

              7.6s (7.2% of total time) in 65 GCs | Peak RSS: 8.19GB | CPU load: 5.31

Produced artifacts:
/Users/gturnquist/src/learning-spring-boot-3rd-edition-code/ch8/target/ch8 (executable)
/Users/gturnquist/src/learning-spring-boot-3rd-edition-code/ch8/target/ch8.build_artifacts.txt (txt)
```

Figure 8.1 – Output from mvnw -Pnative clean native:compile

The resulting artifact is neither an uber JAR file nor an executable JAR file. Instead, it's an executable file for the platform it was built on.

> **Important**
>
> One of Java's most popular features since Day 1 has been its *write once/run anywhere* nature. This works because it's normally compiled into platform-independent **bytecode** and runs inside **JVM**, a virtual machine that can vary from machine to machine. GraalVM sidesteps all of that. The final executable does *NOT* have this run anywhere nature. You can inspect the final app by typing `file target/ch8` in the project's base directory. On my machine, it reads `Mach-O 64-bit executable arm64`.

To run our native application, we just do this:

```
% target/ch8

  .   ____          _            __ _ _
 /\\ / ___'_ __ _ _(_)_ __  __ _ \ \ \ \
( ( )\___ | '_ | '_| | '_ \/ _` | \ \ \ \
 \\/  ___)| |_)| | | | | || (_| |  ) ) ) )
  '  |____| .__|_| |_|_| |_\__, | / / / /
 =========|_|==============|___/=/_/_/_/
 :: Spring Boot ::                (v3.0.0)
......omitted for brevity......
2022-11-06T14:35:40.717-06:00    INFO 12263 ---
[                     main] o.s.b.w.embedded.tomcat.
TomcatWebServer    : Tomcat started on port(s): 8080 (http)
with context path ''
2022-11-06T14:35:40.717-06:00    INFO
12263 --- [                main]
c.s.1.Chapter8Application                          :
Started Chapter8Application in 0.104 seconds (process running
for 0.121)
```

On the last line, we can see that the application started up in 0.104 seconds. For a Java application, that is incredibly fast.

It's possible to get a popup like this:

Figure 8.2 – Native application requesting permission to accept network connections

That was a lot of effort. So, why do all this?

Why do we want GraalVM again?

It's taken a little extra effort to configure our application to work with GraalVM. It also took longer to build the app itself. On top of that, we traded in Java's amazing write-once/run-anywhere flexibility.

Why?

Imagine running 1,000 copies of this application in the cloud. What if our app were to take 20 seconds to start up? 1,000 instances would translate to 20,000 seconds or 5.6 hours.

5.6 hours of billable cloud time.

Adding on an extra 5.6 hours of billable time every time we rolled out a change would start to add up. If we embrace continuous delivery and pushed out every patch commit, our bill could get out of hand. Maybe not from an Ops perspective, but definitely from a billing one!

If, instead, our app launches in 0.1 seconds, like it just did, 1,000 instances would net us just under 17 minutes of cloud time. Whew! That's some cost savings.

Additionally, our application would run in a more efficient memory configuration. So long as our continuous delivery system built the app on the same operating system as our target environment, write-once/run-anywhere isn't an issue.

There is still one lingering issue… what if we don't *HAVE* the target environment on our local build machine? What if we were working in Windows or on a Mac, but our cloud operated using Linux-based Docker containers?

Thankfully, there's a solution for that in the next section!

Baking a Docker container with GraalVM

Earlier in this chapter, we installed GraalVM's OpenJDK distribution and built our native application locally. But that's not the only way, nor is it always the ideal way.

For example, if we plan to run our application on a cloud configuration based on Linux, then building an application locally on a MacBook Pro or a Windows machine won't do.

In *Chapter 7, Releasing an Application with Spring Boot*, we learned how to use `./mvnw spring-boot:build-image` and let a Paketo Buildpack assemble our application into a Docker container. We can use something similar to build a native application inside a Docker container.

Just run the following command:

```
% ./mvnw -Pnative spring-boot:build-image
```

This combines the previous chapter's `spring-boot:build-image` command with the `native` Maven profile.

This process may take even longer than building the native application locally, but the benefit is that, when completed, you will have a fully baked Docker container with a native application in it.

As discussed in the previous chapter, you now have multiple options for running it on your local machine, pushing it to your cloud provider, or releasing the application to Docker Hub.

> **Warning!**
>
> At the time of writing, M1 Macs do *NOT* support this option! If you invoke `./mvnw -Pnative spring-boot:build-image`, it will start the process, but at a certain stage, it will simply hang and never go forward. To stop the process, you *MUST* go into Docker Desktop and kill the Paketo Buildpack that is being used to perform this task. There are overrides to `spring-boot-maven-plugin` that let you plug in alternative Buildpack configurations. If you are native to `spring-boot-starter-parent`, look at its `pom.xml` file from inside your IDE and look for that native profile – you'll see how they configure that plugin with a builder.

Having done all this, there might some lingering confusion out there.

Spring Boot 3.0 versus Spring Boot 2.7 and Spring Native

You may have heard of Spring Native. There have been a lot of blog articles about it. You can even find videos on YouTube talking about using Spring Native (even on my channel!). But as you may have noticed, we haven't mentioned Spring Native until now.

Spring Native was an experimental bridge project built for Spring Boot 2.7. The bits that are in Spring Native have been made first-class citizens in Spring Boot 3 and Spring Framework 6. There is nothing that must be added to your project to compile it into native mode.

We did add GraalVM Native Support from `start.spring.io`, but that was to provide additional support to `spring-boot-maven-plugin`. This brought in `hibernate-enhance-maven-plugin` to help ensure we were building *ALL* the needed metadata to work properly with GraalVM.

But all the AOT processing and metadata management used to make native applications work is in the latest version of the Spring portfolio.

GraalVM and other libraries

The Spring portfolio is being fitted to support GraalVM. Most of their projects support it. But that doesn't mean every third-party library we pick up is supported (yet). As stated in Spring Boot's reference documentation, "*GraalVM native images are an evolving technology and not all libraries provide support.*"

The Spring team is working diligently to not only ensure *ALL* of their modules eventually support native images, but they are also working directly with the GraalVM team to ensure GraalVM itself works properly.

Stay tuned to `spring.io/blog` for future posts concerning improvements for native images.

Summary

In this chapter, we learned how to build native images using GraalVM. This is a faster, more efficient version of our application than we could ever build using the standard JVM. We also learned how to bake native images into a Docker container using Paketo Buildpacks.

In the next chapter, we'll learn how to make Spring Boot apps even more efficient by dipping our toes in the waters of reactive programming.

Part 4:
Scaling an Application with Spring Boot

Sometimes we need to squeeze out more performance from an existing set of servers. While it would be nice to simply buy more servers, there is another way. You will learn how reactive programming can make your Spring Boot application much more efficient.

This part includes the following chapters:

- *Chapter 9, Writing Reactive Web Controllers*
- *Chapter 10, Working with Data Reactively*

9

Writing Reactive Web Controllers

In the previous eight chapters, we gathered up all the key components needed to build a Spring Boot application. We bundled it inside a Docker container and even tweaked it to run in native mode on GraalVM instead of the standard JVM.

But what if, after doing all this, our application still suffered from a lot of idle time? What if our application was burning up our cloud bill due to having to host a huge number of instances just to meet our present needs?

In other words, is there another way to squeeze a lot more efficiency out of the whole thing, without letting go of Spring Boot?

Welcome to Spring Boot and reactive programming!

In this chapter, we'll cover the following topics:

- Discovering exactly what reactive programming is and why we should care
- Creating a reactive Spring Boot application
- Serving data with a reactive GET method
- Consuming incoming data with a reactive POST method
- Serving a reactive template
- Creating hypermedia reactively

Where to find this chapter's code

The code for this chapter can be found at `https://github.com/PacktPublishing/Learning-Spring-Boot-3.0/tree/main/ch9`.

What is reactive and why do we care?

For literally decades, we've seen various constructs meant to help scale applications. This has included thread pools, synchronized code blocks, and other context-switching mechanisms meant to help us run more copies of our code safely.

And, in general, they have all failed.

Don't get me wrong. People run huge systems with some sense of power. But the promises of multithreaded constructs have been tall, their implementation is tricky and frankly hard to get right, and the results have been meager.

People still end up running 10,000 instances of a highly needed service, which can result in a gargantuan monthly bill should we host our application on Azure or AWS.

But what if there were another way? What if the concept of lots of threads and lots of switching were a will-o'-wisp?

Introduction to Reactive

The evidence is in. **Reactive** JavaScript toolkits in the browser, an environment where there is only one thread, have shown incredible ability. That's right – a single-threaded environment can scale and perform with power if approached properly.

We keep using this term, *reactive*. What does it mean?

In this context, we're talking about Reactive Streams. This is from the official documentation:

> *"Reactive Streams is an initiative to provide a standard for asynchronous stream processing with non-blocking back pressure. This encompasses efforts aimed at runtime environments (JVM and JavaScript) as well as network protocols."*
>
> – *Reactive Stream official site.* (`https://www.reactive-streams.org/`)

A key characteristic that has been observed is that fast data streams cannot be allowed to overrun the stream's destination. Reactive Streams addresses this concern by introducing a concept known as backpressure.

Backpressure replaces the traditional publish-subscribe paradigm with a pull-based system. Downstream consumers reach back to publishers and have the power to ask for 1, 10, or however many units to process that it is ready to handle. The mechanism of communication in reactive streams is called **signals**.

Backpressure signals are also baked into the standard in a way that chaining together multiple Reactive Streams components results in backpressure across the entire application.

There is even **RSocket**, a layer 7 network protocol. It is analogous to **HTTP** in that it runs on top of **TCP** (or **WebSockets/Aeron**), and is language agnostic, yet it comes with backpressure built in. Reactive Streams components can communicate over the network in a purely reactive way, with proper control.

What does backpressure allow?

It's not uncommon in traditional systems to be forced to *find the breaking point*. Somewhere in the system is the point where some component is getting overwhelmed. Once this is resolved, the problem simply shifts to the next point, which usually isn't obvious until the primary issue is resolved.

Reactive Stream details

Reactive Streams is a very simple spec, so simple it only has four interfaces: Publisher, Subscriber, Subscription, and Processor:

- **Publisher**: A component that is producing output, whether that's one output or an infinite amount

- **Subscriber**: A component that is receiving from a Publisher

- **Subscription**: Captures the details needed for Subscribers to start consuming content from Publishers

- **Processor**: A component that implements both Subscriber and Publisher

While this is simple, it's frankly too simple. It's recommended to find a toolkit that implements the spec and provides more structures and support to build applications.

The other core thing to understand about Reactive Streams is that it comes with signals. Every time data is handled or actions are taken, they are associated with a signal. Even if there is no data exchange, signals are still handled. This means that there are fundamentally no void methods in reactive programming. That's because even with no data results, there is still the need to send and receive signals.

For the rest of this book, we will use the Spring team's implementation of Reactive Streams, known as **Project Reactor**. It's important to understand that while Project Reactor is produced by the Spring team, Reactor itself doesn't have any Spring dependencies. Reactor is a core dependency picked up by Spring Framework, Spring Boot, and the rest of the Spring portfolio. But it's a toolkit in and of itself.

That means we won't be using Reactive Streams directly but instead Project Reactor's implementation of it. But it's good to understand where it comes from and how it's possible to integrate with other implementations of the spec, such as **RxJava 3**.

Project Reactor is a toolkit that's heavily built on Java 8's functional programming features combined with lambda functions.

Check out the following snippet of code:

```
Flux<String> sample = Flux.just("learning", "spring",
  "boot") //
  .filter(s -> s.contains("spring")) //
  .map(s -> {
    System.out.println(s);
    return s.toUpperCase();
  });
```

This code fragment of Reactor code has some key aspects, as shown here:

- `Flux`: This is Reactor's reactive data flow type of 0 or more data units, each coming at some point in the future.

- `just()`: Reactor's way to create an initial collection of Flux'd elements.

- `filter()`: Similar to Java 8 Stream's `filter()` method, it only allows data elements from the earlier `Flux` through if they satisfy the predicate clause. In this case, does it contain the `"spring"` string?

- `map()`: Similar to Java 8 Stream's `map()` method, it allows you to transform each data element into something else, even a different type. In this scenario, it converts the string into uppercase.

This chunk of code can be described as a flow or a reactive recipe. Each line is captured as a command object in a process known as **assembly**. Something that isn't as obvious is that assembly isn't the same as running things.

When it comes to Reactive Streams, it's important to understand that nothing happens until you **subscribe**.

`onSubscribe` is the first and most important signal in Reactive Streams. It's the indication that a downstream component is ready to start consuming these upstream events.

Once a `Subscription` has been established, `Subscriber` can issue a `request(n)`, asking for n items.

`Publisher` can then start publishing the items through the Subscriber's `onNext` signal. This `Publisher` is free and clear to publish up to (but not exceeding) n invocations of the onNext method.

`Subscriber` can continue to invoke the Subscription's `request` method, asking for more. Alternatively, `Subscriber` can cancel its `Subscription`.

`Publisher` can continue sending content, or it can signal there is no more through the `onComplete` signal.

Now, this all is quite simple and yet… a bit tedious. The recommendation is to let the framework handle this. Application developers are encouraged to write their applications at a higher level, allowing the framework to do all the reactive ceremonies.

And so, in the next section, we'll see how **Spring WebFlux** and Project Reactor make it super simple to build a web controller reactively.

Creating a reactive Spring Boot application

To start writing reactive web apps, we need a completely new application. And to do that, let's revisit our old friend, `https://start.spring.io`.

We'll pick the following settings:

- **Project**: **Maven**
- **Language**: **Java**
- **Spring Boot**: **3.0.0**
- **Group**: `com.springbootlearning.learningspringboot3`
- **Artifact**: `ch9`
- **Name**: `Chapter 9`
- **Description**: `Writing Reactive Web Controllers`
- **Package name**: `com.springbootlearning.learningspringboot3`
- **Packaging**: **Jar**
- **Java**: **17**

With this project metadata selected, we can now start picking our dependencies. Now, instead of adding new things, as we've done in previous chapters, we are starting fresh with the following choices:

- **Spring Reactive Web** (Spring WebFlux)

That's it! That's all we need to get off the ground with building a reactive web application. Later in this chapter and into the next chapter, we'll revisit this to add new modules.

Click **GENERATE** and download the ZIP file; we'll have a tasty little web application with the following key things in the `pom.xml` build file:

- `Spring-boot-starter-webflux`: Spring Boot's starter that pulls in Spring WebFlux, Jackson for JSON serialization/deserialization, and Reactor Netty as our reactive web server
- `spring-boot-starter-test`: Spring Boot's starter for testing, including unconditionally for all projects

- `reactor-test`: Project Reactor's test module with additional tools to help test reactive apps, automatically included with any reactive app

We haven't delved into all the intricacies of reactive programming, but one thing that's required is a web container that isn't stuck on blocking APIs. That's why we have **Reactor Netty**, a Project Reactor library that wraps non-blocking **Netty** with Reactor hooks.

And be aware that testing is vital. That's why the two test modules are also included. And we'll certainly be taking advantage of Reactor's test module later in this chapter and the next one.

But before we can do all that, we need to get familiar with writing a reactive web method, as shown in the next section.

Serving data with a reactive GET method

Web controllers typically do one of two things: serve up data or serve up HTML. To grok the reactive way, let's pick the first since it's much simpler.

In the previous section, we saw a simple usage of Reactor's Flux type. Flux is Reactor's implementation of Publisher and provides a fistful of reactive operators.

We can use it in a web controller like this:

```
@RestController
public class ApiController {
  @GetMapping("/api/employees")
  Flux<Employee> employees() {
    return Flux.just( //
      new Employee("alice", "management"), //
      new Employee("bob", "payroll"));
  }
}
```

This RESTful web controller can be described as follows:

- `@RestController`: Spring Web's annotation to indicate that this controller involves data, not templates

- `@GetMapping`: Spring Web's annotation to map `HTTP GET /api/employees` web calls onto this method

- `Flux<Employee>`: The return type is a `Flux` of `Employee` records

`Flux` is sort of like combining a classic Java `List` with a `Future`. But not really.

Lists can contain multiple items, but a `Flux` doesn't have them all at once. And Flux's aren't consumed through classic iteration or `for` loops. Instead, they come packed with lots of stream-oriented operations, such as `map`, `filter`, `flatMap`, and more.

As for being similar to a `Future`, that's true only in the sense that when a Flux is formed, its contained elements usually don't exist yet, but will instead arrive in the future. But Java's `Future` type, predating Java 8, only has a `get` operation. As mentioned in the previous paragraph, `Flux` has a rich set of operators.

On top of all that, `Flux` has different ways to merge multiple `Flux` instances into a single one, as shown here:

```
Flux<String> a = Flux.just("alpha", "bravo");
Flux<String> b = Flux.just("charlie", "delta");
a.concatWith(b);
a.mergeWith(b);
```

This code can be described as follows:

- a and b: Two pre-loaded `Flux` instances
- `concatWith`: A `Flux` operator that combines a and b into a single `Flux` where all elements of a are emitted before the elements of b
- `mergeWith`: A `Flux` operator that combines a and b into a single `Flux` where the elements are emitted as they come in real time, allowing interleaving between a and b

> **Didn't you pre-load that Flux in the web method with hard-coded data?**
>
> Yes, this example kind of defies the futuristic nature of `Flux` in real-world applications since we don't pre-load a `Flux` using the `just` method. Instead, we're more likely to connect a source of data such as a reactive database or a remote network service. Using Flux's more sophisticated APIs, it's possible to emit entries as they become available into Flux for downstream consumption.

In the web method we first defined, the Flux of data gets handed over to Spring WebFlux, which will then serialize the contents and serve up a JSON output.

It's important to note that all the handling of Reactive Streams signals, including subscription, request, the `onNext` calls, and finally the `onComplete` invocation, are handled by the framework!

It's vital to understand that, in reactive programming, nothing happens until we subscribe. Web calls aren't made and database connections aren't opened. Resources aren't allocated until someone subscribes. The whole system is designed from the ground up to be *lazy*.

But for web methods, we let the framework do the subscribing for us.

Now, let's learn how to craft a web method that consumes data reactively.

Consuming incoming data with a reactive POST method

Any website that will serve up employee records must surely have a way to enter new ones, right? So, let's create a web method that does just that by adding to the `ApiController` class we started in the previous section:

```
@PostMapping("/api/employees")
Mono<Employee> add(@RequestBody Mono<Employee> newEmployee)
  {
    return newEmployee //
      .map(employee -> {
        DATABASE.put(employee.name(), employee);
        return employee;
      });
  }
```

This Spring WebFlux controller has the following details:

- `@PostMapping`: Spring Web's annotation to map HTTP POST `/api/employees` web calls to this method

- `@RequestBody`: This annotation tells Spring Web to deserialize the incoming HTTP request body into an `Employee` data type

- `Mono<Employee>`: Reactor's alternative to `Flux` for a single item

- `DATABASE`: A temporary data store (a Java `Map`)

The incoming data is wrapped inside a Reactor Mono. This is the single-item counterpart to a Reactor `Flux`. By mapping over it, we can access its contents. Reactor Mono also supports many operators, as `Flux` does.

While we can transform the contents, in this situation, we are simply storing the content in our `DATABASE` and then returning it with no change.

---

map versus flatMap

We've now seen map used twice in the initial chunk of code at the start of this chapter and also in this latest method. Mapping is a one-to-one operation. If we were to map over a `Flux` with 10 items, the new `Flux` would also have 10 items. What if a single item, such as a string, were mapped into a list of its letters? The transformed type would be a list of lists. In many such situations, we'd like to collapse the nesting and simply have a new `Flux` with all the letters. This is flattening! `flatMap` is the same thing, just done in one step!

Scaling applications with Project Reactor

So, how exactly does Project Reactor scale our application? So far, we've seen how Project Reactor gives us a functional programming style. But it may not be clear exactly where scalability comes into play.

That's because Project Reactor handles two key things seamlessly under the covers. The first is that every step in these little flows or recipes isn't directly carried out. Instead, every time we write a map, a filter, or some other Reactor operator, we are *assembling* things. This isn't when execution happens.

Each of these operations assembles a tiny command object with all the details needed to carry out our actions. For example, the statement in the previous code block where we store the value in DATABASE and then return the value is all wrapped inside a Java 8 **lambda** function. There is no requirement that when the controller method is invoked, this inner lambda function is called at the same time.

Project Reactor gathers all these command objects representing our actions and stacks them on internal work queues. It then delegates execution to its built-in Scheduler. This makes it possible for Reactor to decide *exactly* how to carry things out. There are various Schedulers to choose from, including a single thread to a thread pool, to a Java ExecutorService to a sophisticated bounded elastic scheduler.

Your Scheduler of choice works through its backlog of work as system resources become available. By having every single step of a Reactor flow act in a lazy, non-blocking way, anytime there is an I/O-bound delay, the current thread isn't held up waiting for a response. Instead, Scheduler goes back to this internal queue of work and picks a different task to carry out. This is known as **work stealing** and makes it possible for traditional latency issues to turn into opportunities to complete other work, resulting in better overall throughput.

As Spring Data team leader Mark Paluch once said, *"Reactive programming is based on reacting to resource availability."*

Earlier, we mentioned that Reactor does two things. The second thing is that instead of having a giant thread pool with 200 threads, it defaults to a Scheduler using a pool with one thread per core.

Quick history on Java concurrent programming

In the early days of Java concurrent programming, people would create giant pools of threads. But we learned that when you have more threads than cores, context switching becomes expensive.

On top of that, Java has many breaking APIs baked into its core. While we were given tools since the early days such as synchronized methods and blocks, along with locks and semaphores, it's really hard to do this both effectively and correctly.

It was easy to either A) do it right but not increase throughput, B) improve throughput yet introduce deadlocks, or C) introduce deadlocks *and* not improve throughput. And these tactics often require rewriting the application in a non-intuitive way.

One thread per core when combined with lazy, non-blocking work stealing can be *MUCH* more efficient. Of course, coding with Project Reactor isn't invisible. There is a programming style to incorporate, but since it's heavily tilted toward the same style of programming as Java 8 Streams, which has been widely adopted, it's not the big ask that early-days Java concurrent programming was.

This is also the reason that absolutely every part of the application needs to be written this way. Imagine a 4-core machine that has only 4 Reactor threads. If one of these threads ran into blocking code and were forced to wait, it would clobber 25% of the total throughput.

This is the reason that blocking APIs found in places such as JDBC, JPA, JMS, and servlets is a profound issue for reactive programming.

All these specifications were built on top of blocking paradigms, so they aren't suitable for reactive applications, which we'll explore in more detail in the next chapter, *Working with Data Reactively*.

In the meantime, let's learn how to implement a reactive template.

Serving a reactive template

So far, we've built a reactive controller that serves up some serialized JSON. But most websites need to render HTML. And this leads us to templates.

Since we're talking about reactive programming, it makes sense to pick a templating engine that doesn't block. So, for this chapter, we'll be using **Thymeleaf**.

To get going, first, we need to update the application we started building at the beginning of this chapter. To do that, let's revisit `https://start.spring.io`.

We've done this dance in previous chapters. Instead of making an entirely new project and starting over (ugh!), instead, we will enter all the same project metadata shown earlier in this chapter in the *Creating a reactive Spring Boot application* section.

This time, enter the following dependencies:

- **Spring Reactive Web**
- **Thymeleaf**

Now, instead of using **GENERATE** like we did last time, hit the **EXPLORE** button. This will cause the web page to serve up an online preview of this barebones project, showing us the `pom.xml` build file.

Everything should be the same as the `pom.xml` file we downloaded earlier with one difference: that new dependency for `spring-boot-starter-thymeleaf`. All we need to do is this:

1. Highlight that Maven dependency.
2. Copy it to the clipboard.
3. Paste it into our IDE's `pom.xml` file.

This extra Spring Boot starter will download Thymeleaf, a templating engine that not only integrates nicely with Spring Boot but also comes loaded with reactive support. This sets us up to write a reactive web controller for templates, as shown in the next section.

Creating a reactive web controller

The next step is to create a web controller focused on serving templates. To do that, create a new file called HomeController.java and add the following code:

```java
@Controller
public class HomeController {
  @GetMapping("/")
  public Mono<Rendering> index() {
    return Flux.fromIterable(DATABASE.values()) //
      .collectList() //
      .map(employees -> Rendering //
        .view("index") //
        .modelAttribute("employees", employees) //
        .build());
  }
}
```

There's a lot in this controller method, so let's take it apart:

- @Controller: Spring Web's annotation to indicate this class contains web methods that render templates.

- @GetMapping: Spring Web's annotation to map GET / web calls onto this method.

- Mono<Rendering>: Mono is Reactor's single-valued reactive type. Rendering is Spring WebFlux's value type that allows us to pass both the name of the view to render along with model attributes.

- Flux.fromIterable(): This static helper method lets us wrap any Java Iterable and then use our reactive APIs.

- DATABASE.values(): This is a temporary data source until we get to *Chapter 10, Working with Data Reactively*.

- collectList(): This Flux method lets us gather a stream of items into Mono<List<Employee>>.

- map(): This operation lets us access the list inside that Mono where we then transform it into a Rendering. The name of the view we wish to render is "index". We also load up the model "employees" attribute with the values found inside this Mono.

- build(): Rendering is a builder construct, so this is the step that transforms all the pieces into a final, immutable instance. It's important to understand that when inside the map() operation, the output is a Mono<Rendering>.

There are some other aspects of this web method that are important to understand.

First of all, the map() operation at the end of the chain is meant to transform the type that's inside Mono. In this case, it converts List<Employee> into a Rendering, while keeping everything inside this Mono. It does this by unpacking the original Mono<List<Employee>> and using the results to create a brand new Mono<Rendering>.

> **Functional programming 101**
>
> It's basic functional programming to have some container such as a Flux or a Mono and you map over what's inside, all the time, keeping it inside the functional Flux or Mono. You don't have to worry about making new Mono instances. The Reactor APIs are designed to handle that for you. You focus on transforming the data along the way. So long as things remain smoothly encased in these reactive container types, the framework will properly unpack them at the right time and render them properly.

The other thing that is important to recognize is that we aren't using a *real* data source. This is canned data stored in a basic Java Map. That's why it may seem a bit strange to have wrapped a Java list of Employee objects into a Flux using fromIterable, only to extract them back out using collectList.

This is to illustrate the real-world situation we often face of being handed an Iterable collection. The proper course of action is what's shown in the code: wrap it into a Flux and then execute our various transformations and filters, until we hand it off to the web handlers of Spring WebFlux to be rendered with Thymeleaf.

The one thing remaining is to code that template using Thymeleaf!

Crafting a Thymeleaf template

The final step in building a template-based solution is to create a new file, index.html, underneath src/main/resources/templates, as shown here:

```
<html xmlns:th="http://www.thymeleaf.org">
<head>
  <title>Writing a Reactive Web Controller</title>
</head>
```

```
<body>
<h2>Employees</h2>
<ul>
  <li th:each="employee : ${employees}">
    <div th:text=
      "${employee.name + ' (' + employee.role + ')'}">
    </div>
  </li>
</ul>
</body>
</html>
```

This template can be described as follows:

- `xmlns:th=http://www.thymeleaf.org`: This XML namespace allows us to use Thymeleaf's HTML processing directives.

- `th:each`: Thymeleaf's `for-each` operator, giving us a `<li>` node for every entry in the `employees` model attribute. In each node, `employee` is the stand-in variable.

- `th:text`: Thymeleaf's directive to insert text into the node. In this situation, we are concatenating two attributes from the employee record with strings.

Something else probably not visible is that *ALL* of the HTML tags in this template are closed. That is, no tag can be left open due to Thymeleaf's DOM-based parser. Most HTML tags have an opening and closing tag, but some do not, such as `<IMG>`. When using such tags in Thymeleaf, we must close them, either with a corresponding `</IMG>` tag or by using the `<IMG/>` shortcut.

> **Thymeleaf's pros and cons**
>
> If it's any consolation, anytime I use Thymeleaf, I must have their reference page open for me to look at. If I coded Thymeleaf every day, perhaps I wouldn't need this. Despite this aspect, Thymeleaf is quite powerful. There are extensions such as support for Spring Security and the ability to write security checks into your templates, allowing certain elements to be rendered (or not) based on the user's credentials and authorizations. Thymeleaf on the whole can probably do anything you need when it comes to crafting HTML at the expense of having to take on its notation.

If you fire up our reactive application and navigate to `http://localhost:8080`, you'll see a nice little rendered web page:

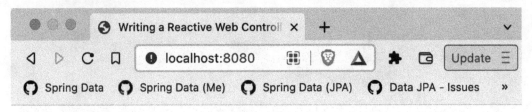

Employees

- Frodo Baggins (ring bearer)
- Samwise Gamgee (gardener)
- Bilbo Baggins (burglar)

Figure 9.1 – Reactive template rendered with Thymeleaf

This wouldn't be much of a website if we didn't include the ability to add to our employee database. To do that, we need to enter the world of form-binding.

In general, to POST a new object, we must first provide an empty object during the GET part of the operation, so we need to update our index method, as shown here:

```
@GetMapping("/")
Mono<Rendering> index() {
  return Flux.fromIterable(DATABASE.values()) //
    .collectList() //
    .map(employees -> Rendering //
      .view("index") //
      .modelAttribute("employees", employees) //
      .modelAttribute("newEmployee", new Employee("", ""))
      .build());
}
```

This code is the same as the previous version except for the highlighted line. It introduces a new model attribute, newEmployee, containing an empty Employee object. This is all that's needed to start crafting an HTML form with Thymeleaf.

In the index.html template we created in the previous section, we need to add the following:

```
<form th:action="@{/new-employee}" th:object=
  "${newEmployee}" method="post">
    <input type="text" th:field="*{name}" />
```

```
    <input type="text" th:field="*{role}" />
    <input type="submit" />
</form>
```

This Thymeleaf template can be described as follows:

- th:action: Thymeleaf's directive to form a URL to a route that we'll code further down to process new Employee records

- th:object: Thymeleaf's directive to bind this HTML form to the newEmployee record that was provided as a model attribute in our updated index method

- th:field="*{name}": Thymeleaf's directive to connect the first <input> to the Employee record's name

- th:field="*{role}": Thymeleaf's directive to connect the second <input> to the Employee record's role

The rest is standard HTML 5 <form>, which we won't go into. The parts expounded upon are the glue needed to hook HTML form processing into Spring WebFlux.

The last step is to code a POST handler in our HomeController, as shown here:

```
@PostMapping("/new-employee")
Mono<String> newEmployee(@ModelAttribute Mono<Employee>
  newEmployee) {
    return newEmployee //
      .map(employee -> {
        DATABASE.put(employee.name(), employee);
        return "redirect:/";
      });
}
```

This operation can be described as follows:

- @PostMapping: Spring Web's annotation to map POST /new-employee web calls to this method.

- @ModelAttribute: Spring Web's annotation to signal that this method is meant to consume HTML forms (versus something such as an application/json request body).

- Mono<Employee>: This is the inbound data from the HTML form, wrapped in a Reactor type.

- map(): By mapping over the incoming results, we can extract the data, store it in DATABASE, and transform it into an HTTP redirect operation back to /. This results in a Mono<String> method return type.

Once again, the entire action taken in this method is a Reactor flow that starts with the incoming data and results in a transformation to an outgoing action. Reactor-based programming often has this style compared to classic, imperative programming where we fiddle with intermediate variables.

If we run everything again, we will see this:

Figure 9.2 – Entering a new employee into the form

On this page, the user is entering a new employee record. Once they hit **Submit**, the POST processor kicks in, stores the user, then directs the web page back to the home page.

This causes an updated version of DATABASE to be retrieved, as shown here:

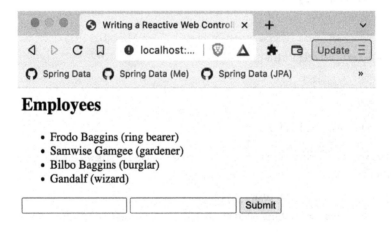

Figure 9.3 – After hitting Submit, the page redirects back to home

The newly entered employee now shows up on the web page.

> **Is Spring WebFlux worth it?**
>
> This type of flow can indeed feel a little daunting, even challenging at first. But over time, it becomes a consistent style that often forces us into thinking about every step. For a standard web app, you can question whether or not this is worth it. But if you have an app that is requiring you to run hundreds if not thousands of copies, and your cloud bill is going through the roof, then there may be a valid, economical reason to consider using Spring WebFlux. Check out my article, *Why Reactive Streams are the SECRET to CUTTING your monthly cloud bill*, at `https://springbootlearning.com/cloud-bill`.

We could delve further into all the various ways to mix Spring WebFlux and Thymeleaf, but this is *Learning Spring Boot 3.0*, not *Learning Thymeleaf 3.1*. And with that, let's look into one other valuable ability… crafting hypermedia-powered APIs.

Creating hypermedia reactively

At the beginning of this chapter, we crafted a very simple API. It served some pretty basic JSON content. One thing that was missing from such a bare API was any controls.

Hypermedia is the term used to refer to both content and metadata being served by an API; this content and metadata indicate what can be done with the data or how to find other related data.

Hypermedia is something we see every day. At least on web pages. This includes the navigation links to other pages, links to CSS stylesheets, and links to effect change. This is quite common. When we order some product from Amazon, we aren't required to provide the link to make it happen. The web page gives it to us.

Hypermedia in JSON is simply the same concept but applied to APIs instead of visual web pages.

And this is easy if we add **Spring HATEOAS** to our application!

> **Spring Boot Starter HATEOAS versus Spring HATEOAS**
>
> If you go to `start.spring.io` and ask for **Spring HATEOAS**, you will have `spring-boot-starter-hateoas` added to your application. But this version is wrong when using Spring WebFlux. For the longest time, Spring HATEOAS only supported Spring MVC, but about 4 years ago, yours truly added WebFlux support. Unfortunately, the Spring Boot Starter HATEOAS module pulls in Spring MVC and Apache Tomcat support, which is the opposite of what we want for a Spring WebFlux application running on top of Reactor Netty. The simplest approach is to just add Spring HATEOAS directly, as shown here.

To add Spring HATEOAS to a reactive application, just add the following dependency:

```
<dependency>
    <groupId>org.springframework.hateoas</groupId>
    <artifactId>spring-hateoas</artifactId>
</dependency>
```

And once again, thanks to Spring Boot's dependency management, there is no need to specify the version number.

With that in place, we can start building a hypermedia-powered API. First, create a class named `HypermediaController.java`, like this:

```
@RestController
@EnableHypermediaSupport(type = HAL)
public class HypermediaController {
}
```

This barebones controller class can be described as follows:

- `@RestController`: Spring Web's annotation to mark this controller as focused on JSON serializing instead of template rendering
- `@EnableHypermediaSupport`: Spring HATEOAS's annotation to activate hypermedia support – in this case, HAL support

If we had used Spring Boot Starter HATEOAS, the HAL support would have been activated automatically. But since we're manually plugging in Spring HATEOAS, we must activate it ourselves.

Important

The `@EnableHypermediaSupport` annotation only has to be used once. We happen to be putting on our hypermedia controller for brevity in this book. In a *real* application, it may be preferable to put it in the same class that has the `@SpringBootApplication` annotation.

With all this in place, let's start by building a hypermedia endpoint for a single-item resource, one employee, as shown here:

```
@GetMapping("/hypermedia/employees/{key}")
Mono<EntityModel<Employee>> employee(@PathVariable String
    key) {
        Mono<Link> selfLink = linkTo( //
            methodOn(HypermediaController.class) //
```

```
                    .employee(key)) //
                       .withSelfRel() //
                       .toMono();

  Mono<Link> aggregateRoot = linkTo( //
    methodOn(HypermediaController.class) //
       .employees()) //
          .withRel(LinkRelation.of("employees"))//
          .toMono();

  Mono<Tuple2<Link, Link>> links = Mono.zip(selfLink,
    aggregateRoot);

  return links.map(objects ->
    EntityModel.of(DATABASE.get(key), objects.getT1(),
      objects.getT2()));
}
```

This implementation for the single-item `Employee` is loaded with content, so let's take it apart:

- `@GetMapping`: Spring Web's annotation to indicate that this will serve `HTTP GET /
  hypermedia/employee/{key}` methods.

- The return type is `Mono<EntityModel<Employee>>`. `EntityModel` is Spring
  HATEOAS's container for an object that includes links. And we've already seen how `Mono` is
  Reactor's wrapper for reactive programming.

- `linkTo()`: Spring HATEOAS's static helper function to extract a link from a Spring WebFlux
  method invocation.

- `methodOn()`: Spring HATEOAS's static helper function to perform a dummy invocation of
  a controller's web method to gather information for building links. In the first usage, we are
  pointing at the `employee(String key)` method of `HypermediaController`. In the
  second usage, we are pointing at the `employees()` method of `HypermediaController`
  (not yet written).

- `withSelfRel()`: Spring HATEOAS's method to label `selfLink` with a `self` **hypermedia
  relation** (which we'll see shortly).

- `withRel(LinkRelation.of("employees"))`: Spring HATEOAS's method to apply
  an arbitrary `employees` hypermedia relation.

- `toMono()`: Spring HATEOAS's method to take all the link-building settings and turn them into a `Mono<Link>`.

- `Mono.zip()`: Reactor's operator to combine two `Mono` operations and process the results when both are complete. There are other utilities for bigger sets, but waiting for two is so common that `zip()` is kind of a shortcut.

- `links.map()`: We map over `Tuple2` of the `Mono<Link>` object, extracting the links and bundling them with the fetched employee into a Spring HATEOAS `EntityModel` object.

So what does Spring HATEOAS do?

It combines data (which we've been using all along) with hyperlinks. Hyperlinks are represented using Spring HATEOAS's `Link` type. The toolkit is filled with operations to ease the creation of `Link` objects and merge them with data. The previous block of code showed how to extract links from a Spring WebFlux method.

For Spring HATEOAS to render this merging of data and links, we need to have any hypermedia-powered endpoint return a Spring HATEOAS `RepresentationModel` object or one of its subtypes. The list isn't long and is shown here:

- `RepresentationModel`: The core type for data and links. One option for a single-item hypermedia type is to extend this class and merge our business values with it.

- `EntityModel<T>`: The generic extension of `RepresentationModel`. Another option is to inject our business object into its static constructor methods. This lets us keep links and business logic separated from each other.

- `CollectionModel<T>`: The generic extension of `RepresentationModel`. It represents a collection of `T` objects instead of just one.

- `PagedModel<T>`: The extension of `CollectionModel` that represents a page of hypermedia-aware objects.

It's important to understand that a single-item hypermedia-aware object can have one set of links while a collection of hypermedia-aware objects can have a different set of links. To properly represent a rich collection of hypermedia-aware objects, it can be captured as a `CollectionModel<EntityModel<T>>`.

This would imply that the entire collection may have one set of links, such as a link to the aggregate root. And each entry of the collection may have a custom link pointing to its single-item resource method while they all have a link back to the aggregate root.

To better see this, let's implement that aggregate root – the hypermedia-aware end mentioned in the previous code block:

```
@GetMapping("/hypermedia/employees")
Mono<CollectionModel<EntityModel<Employee>>> employees() {
Mono<Link> selfLink = linkTo( //
  methodOn(HypermediaController.class) //
    .employees()) //
      .withSelfRel() //
      .toMono();

return selfLink //
  .flatMap(self -> Flux.fromIterable(DATABASE.keySet()) //
    .flatMap(key -> employee(key)) //
    .collectList() //
    .map(entityModels -> CollectionModel.of(entityModels,
      self)));
}
```

Part of this method should look very similar to the previous code block. Let's focus on the differences:

- `@GetMapping`: This method maps `GET /hypermedia/employees` to this method, the aggregate root.

- `selfLink` in this method points to this method, which is a fixed endpoint.

- We `flatMap()` over `selfLink`, and then extract every entry from `DATABASE`, leveraging the `employee(String key)` method to convert each entry into an `EntityModel<Employee>` with single-item links.

- We use `collectList()` to bundle all this into a `Mono<List<EntityModel<Employee>>>`.

- Finally, we map over it, converting it into `Mono<CollectionModel<EntityModel <Employee>>>` with the aggregate root's `selfLink` wired in.

If this appears a lot more complex than previous methods in this chapter (or previous chapters), that's because it is. But hooking web controller methods directly into the rendering of the hypermedia output ensures that future tweaks and adjustments to our methods will adjust properly.

If we run the application, we will easily see where this leads us:

```
% curl -v localhost:8080/hypermedia/employees | jq
{
    "_embedded": {
      "employeeList": [
        {
          "name": "Frodo Baggins",
          "role": "ring bearer",
          "_links": {
            "self": {
              "href": "http://localhost:8080/hypermedia/
                       employees/Frodo%20Baggins"
            },
            "employees": {
              "href": "http://localhost:8080/hypermedia/
                       employees"
            }
          }
        },
        {
          "name": "Samwise Gamgee",
          "role": "gardener",
          "_links": {
            "self": {
              "href": "http://localhost:8080/hypermedia/
                       employees/Samwise%20Gamgee"
            },
            "employees": {
              "href": "http://localhost:8080/hypermedia/
                       employees"
            }
          }
        },
        {
          "name": "Bilbo Baggins",
          "role": "burglar",
```

```
        "_links": {
          "self": {
            "href": "http://localhost:8080/hypermedia/
                    employees/Bilbo%20Baggis"
          },
          "employees": {
            "href": "http://localhost:8080/hypermedia/
                    employees"
          }
        }
      }
    }
  ]
},
"_links": {
  "self": {
    "href": "http://localhost:8080/hypermedia/
            employees"
  }
}
}
```

That's a lot of information. Let's highlight some key parts:

- _link: HAL's format for showing a hypermedia link. It contains a **link relation** (for example, self) and an href (for example, http://localhost:8080/hypermedia/employees)

- The collection's self-link is at the bottom, while the two single-item Employee objects each have a self pointed at itself as well as an employees link pointed to the aggregate root

It's left as an exercise for you to explore the single-item HAL output.

What's a "self" link?

In hypermedia, just about any representation will include what's called a "self" link. It's the *this* concept. Essentially, it's a pointer to the current record. It's important to understand the context. For example, the HAL output shown previously has three different self links. Only the last one is this document's self. The others are the canonical links to look up that individual record. And because links are essentially opaque, you can use these links to navigate to those records.

What is the point of doing all this?

Imagine a system where not only do we have employee-based data, but also other various operations. For example, we could build up a whole host of functions such as `takePTO`, `fileExpenseReport`, and `contactManager`. Anything, really.

Building up a collection of links that cross various systems and appear and disappear based on when they're valid makes it possible to show/hide buttons on web applications based on whether they are relevant.

> **To hypermedia or not to hypermedia, that is the question**
>
> Hypermedia makes it possible to dynamically connect users with related operations and relevant data. Since there's not enough room in this chapter to delve into all the pros and cons of hypermedia, check out *Spring Data REST: Data meets Hypermedia* at `https://springbootlearning.com/hypermedia`, where you can watch me and my teammate Roy Clarkson go into the details of hypermedia and Spring.

Summary

In this chapter, we learned several key skills, including creating a reactive application using Project Reactor, rolling out a reactive web method to both serve and consume JSON, and how to leverage Thymeleaf to reactively generate HTML and consume an HTML form. We even used Spring HATEOAS to reactively generate a hypermedia-aware API.

All these features are the building blocks of web applications. While we used a Java 8 functional style to chain things together, we were able to reuse the same Spring Web annotations we've used throughout this book.

And by using Reactor's style, a paradigm quite similar to Java 8 Streams, we can potentially have a much more efficient application.

That concludes this chapter! In the next chapter, we will wrap things up by showing you how to reactively work with real data.

10

Working with Data Reactively

In the previous chapter, we learned how to write a reactive web controller using Spring WebFlux. We loaded it with canned data and used a reactive templating engine, Thymeleaf, to create an HTML frontend. We also created a reactive API with pure JSON and then with hypermedia using Spring HATEOAS. However, we had to use canned data. That's because we didn't have a reactive data store on hand, an issue we will solve in this chapter.

In this chapter, we'll be covering the following topics:

- Learning what it means to fetch data reactively
- Picking a reactive data store
- Creating a reactive data repository
- Trying out R2DBC

> **Where to find this chapter's code**
> The code for this chapter can be found in this repository: `https://github.com/PacktPublishing/Learning-Spring-Boot-3.0/tree/main/ch10`.

Learning what it means to fetch data reactively

In the previous chapter, we covered a lot of the basics needed to build a web page reactively, except we were missing a vital ingredient: real data.

Real data comes from databases.

There are few applications that don't use a database to manage their data. And in this era of e-commerce sites serving global communities, there has never been a broader choice of various types of databases, be they relational, key-value, document, or whatever.

This can make it tricky to pick the right one for our needs. It's even harder considering that even a database needs to be accessed *reactively*.

That's right. If we don't access our database using the same reactive tactics introduced in the previous chapter, all of our efforts would be for naught. To repeat a key point: *all parts of the system must be reactive*. Otherwise, we risk a blocking call tying up a thread and clobbering our throughput.

Project Reactor's default thread pool size is the number of cores on the operating machine. That's because we've seen that context switching is expensive. By having no more threads than cores, we're guaranteed to never have to suspend a thread, save its state, activate another thread, and restore its state.

By taking such an expensive operation off the table, reactive applications can instead focus on the more effective tactic of simply going back to Reactor's runtime for the next *task* (a.k.a. **work stealing**). However, this is only possible when we use Reactor's Mono and Flux types along with their various operators.

If we ever invoke some blocking call on a remote database, the entire thread will halt, waiting for an answer. Imagine a four-core machine having one of its cores blocked like that. A four-core system suddenly only using three cores would show an instant 25% drop in throughput.

This is the reason that database systems across the spectrum are implementing alternative drivers that use the Reactive Streams specification: MongoDB, Neo4j, Apache Cassandra, Redis, and more.

But what, exactly, does a reactive driver look like? Database drivers handle the process of opening a connection to a database, parsing queries and turning them into commands, and finally, shepherding the results back to the caller. Reactive Streams-based programming has been growing in popularity, which has motivated these various vendors to build reactive drivers.

But one area that is stuck is JDBC.

When it comes to Java, all toolkits, drivers, and strategies go through JDBC to talk to a relational database. jOOQ, JPA, MyBatis, and QueryDSL all use JDBC under the hood. And because JDBC is blocking, it simply won't work in a reactive system.

> **Are you sure?**
>
> People have asked in various ways why can't we just carve out a JDBC thread pool and put a reactor-friendly proxy in front of it. The truth is, while each incoming request could be dispatched to a thread pool, you run into the risk of hitting the pool's limit. At that point, the next reactive call would be blocked, waiting for a thread to free up, effectively clobbering the whole system. The point of reactive systems is to NOT block but instead yield so that other work can be done. A thread pool simply delays the inevitable while costing you the overhead of context switching. A database driver, right up to the point of talking to the database engine itself, needs to speak Reactive Streams or it won't cut it.

JDBC being a spec and not simply a driver makes this all but impossible. But there's hope, as we'll see in the next section.

Picking a reactive data store

Realizing that JDBC wouldn't be able to change enough to support Reactive Streams and needing to serve the growing community of Spring users that wanted to go reactive, the Spring team embarked upon a new solution in 2018. They drafted the **Reactive Relational Database Connectivity (R2DBC)** specification.

R2DBC as a specification reached 1.0 in early 2022, and for the rest of this chapter, we'll be using it to build a reactive relational data story.

> **R2DBC? Tell me more!**
>
> If you want a little more detail about R2DBC, please check out former Spring Data team leader Oliver Drotbohm's keynote presentation from the 2018 SpringOne conference at `https://springbootlearning.com/r2dbc-2018`.

Since we want something super simple, we can use H2 as our relational database of choice. H2 is an in-memory, embeddable database. It's frequently used for test purposes, but we can use it as a *production* application for now.

Along with H2, we'll also use Spring Data R2DBC. To get both of these, let's visit our old friend, `https://start.spring.io`. If we pick the same version of Spring Boot as the previous chapter and plug in the same metadata, we can then choose the following dependencies:

- **H2 Database**

- **Spring Data R2DBC**

If we then click on the **EXPLORE** button and scroll about halfway down the pom.xml file, we should see the following three entries:

```
<dependency>
    <groupId>org.springframework.boot</groupId>
    <artifactId>spring-boot-starter-data-r2dbc</artifactId>
</dependency>
<dependency>
    <groupId>com.h2database</groupId>
    <artifactId>h2</artifactId>
    <scope>runtime</scope>
</dependency>
<dependency>
    <groupId>io.r2dbc</groupId>
    <artifactId>r2dbc-h2</artifactId>
    <scope>runtime</scope>
</dependency>
```

These three dependencies can be described as follows:

- `spring-boot-starter-data-r2dbc`: Spring Boot's starter for Spring Data R2DBC
- `h2`: The third-party embeddable database
- `r2dbc-h2`: The Spring team's R2DBC driver for H2

It's important to understand that R2DBC is very low-level. It's fundamentally aimed at making it easy for database driver authors to implement. Certain aspects of JDBC as a driver interface were compromised to make it easier to consume by applications. R2DBC has sought to remedy this. The consequence is that having an application talk directly through R2DBC is actually quite cumbersome.

That's why it's recommended to use a toolkit. In this case, we'll be using Spring Data R2DBC, but you can pick another that you'd prefer, such as Spring Framework's `DatabaseClient` or another third-party one.

With our tools set up, it's time to start building a reactive data repository!

Creating a reactive data repository

Earlier in *Chapter 3*, *Querying for Data with Spring Boot*, we built an easy-to-read data repository by extending `JpaRepository` from Spring Data JPA. For Spring Data R2DBC, let's write something like this:

```
public interface EmployeeRepository extends //
    ReactiveCrudRepository<Employee, Long> {}
```

This code can be described as follows:

- `EmployeeRepository`: The name for our Spring Data repository.
- `ReactiveCrudRepository`: Spring Data Commons' base interface for any reactive repository. Note that this isn't specific to R2DBC but instead for ANY reactive Spring Data module.
- `Employee`: The domain type for this repository (which we'll code further in the chapter).
- `Long`: The primary key's type.

In the previous chapter, we scratched together an `Employee` domain type using a **Java 17 record**. However, to interact with our database, we need something a little more detailed than that, so let's craft the following:

```
public class Employee {
    private @Id Long id;
    private String name;
```

```
    private String role;

    public Employee(String name, String role) {
      this.name = name;
      this.role = role;
  }
    // getters, setters, equals, hashCode, and toString
      omitted for brevity
  }
```

This code can be described as follows:

- Employee: Our domain's type, as required in the EmployeeRepository declaration.

- @Id: Spring Data Commons' annotation to denote the field that contains the **primary key**. Note that this is NOT JPA's jakarta.persistence.Id annotation but instead a Spring Data-specific annotation.

- name and role: The two other fields we'll be using.

The rest of this domain type's methods can be generated by any modern IDE using its utilities.

With all this, we're ready to start speaking R2DBC!

Trying out R2DBC

Before we can fetch any data, we have to load some data. While this is normally something our DBA deals with, for this chapter we'll just have to do it ourselves. To do that, we need to create a Spring component that will automatically kick off once our application is up. Create a new class named Startup and add the following code:

```
@Configuration
public class Startup {
  @Bean
  CommandLineRunner initDatabase(R2dbcEntityTemplate
    template) {
      return args -> {
        // Coming soon!
      }
    }
}
```

This code can be described as follows:

- `@Configuration`: Spring's annotation to flag this class as a collection of bean definitions, needed to autoconfigure our application

- `@Bean`: Spring's annotation to turn this method into a Spring bean, added to the application context

- `CommandLineRunner`: Spring Boot's functional interface for an object that is automatically executed once the application is started

- `R2dbcEntityTemplate`: Inject a copy of this Spring Data R2DBC bean so we can load a little test data

- `args -> {}`: A Java 8 lambda function that is coerced into `CommandLineRunner`

What do we put inside this Java 8 lambda function? Well, for Spring Data R2DBC, we need to define the schema ourselves. If it's not already defined externally (which it isn't for this chapter), we need to write something like this:

```
template.getDatabaseClient() //
  .sql("CREATE TABLE EMPLOYEE (id IDENTITY NOT NULL
    PRIMARY KEY , name VARCHAR(255), role
      VARCHAR(255))") //
  .fetch() //
  .rowsUpdated() //
  .as(StepVerifier::create) //
  .expectNextCount(1) //
  .verifyComplete();
```

This chunk of code can be described as follows:

- `template.getDatabaseClient()`: For pure SQL, we need access to the underlying `DatabaseClient` from Spring Framework's R2DBC module that does all the work.

- `sql()`: A method to feed a SQL `CREATE TABLE` operation. This uses H2's dialect to create an `EMPLOYEE` table with a self-incrementing `id` field.

- `fetch()`: The operation that carries out the SQL statement.

- `rowsUpdate()`: Gets back the number of rows affected so we can verify that it worked.

- `as(StepVerifier::create)`: Reactor Test's operator to convert this whole reactive flow into `StepVerifier`. `StepVerifier` is another way to conveniently force a reactive flow to be executed.

- `expectNextCount(1)`: Verifies whether we got back one row, indicating the operation worked.

- `verifyComplete()`: Ensures we received a Reactive Streams `onComplete` signal.

The preceding method lets us run a little bit of SQL code in order to create our barebones schema. And it may have gotten a little confusing when, halfway down, we pivoted into Reactor Test's `StepVerifier`.

`StepVerifier` is very handy for testing out reactor flows, but it also affords us a useful means to force small reactor flows while allowing us to see the results when needed. The only problem is that we can't use this because, by default, `reactor-test` is test-scoped when we use the Spring Initializr. To make the preceding code work, we must go into our `pom.xml` file and remove the `<scope>test</scope>` line. Then, refresh the project, and it should work!

Now, with this in place, let's load some data.

Loading data with R2dbcEntityTemplate

So far, we've set up the schema for our `Employee` domain type. With that in place, we're ready to add some more `R2dbcEntityTemplate` calls inside that `initDatabase()` CommandLineRunner:

```
template.insert(Employee.class) //
    .using(new Employee("Frodo Baggins", "ring bearer"))
    .as(StepVerifier::create) //
    .expectNextCount(1) //
    .verifyComplete();

template.insert(Employee.class) //
    .using(new Employee("Samwise Gamgee", "gardener")) //
    .as(StepVerifier::create) //
    .expectNextCount(1) //
    .verifyComplete();

template.insert(Employee.class) //
    .using(new Employee("Bilbo Baggins", "burglar")) //
    .as(StepVerifier::create) //
    .expectNextCount(1) //
    .verifyComplete();
```

These three calls all have the same pattern. Each can be described as follows:

- `insert(Employee.class)`: Defines an insert operation. By providing a type parameter, the subsequent operations are typesafe.

- `using(new Employee(...))`: Here is where the actual data is provided.

- `as(StepVerifier::create)`: The same pattern of using Reactor Test to force the execution of our reactive flow.

- `expectNextCount(1)`: For a single insert, we are expecting a single response.

- `verifyComplete()`: Verifies whether we received the `onComplete` signal.

The `insert()` operation actually returns `Mono<Employee>`. It would be possible to inspect the results, even getting hold of the newly minted `id` value. But since we're just loading data, all we want is to confirm that it worked.

In the next section, we'll see how to hook up our reactive data supply to an API controller.

Returning data reactively to an API controller

The heavy work has been done. From here on in, we can tap into what we learned in the previous chapter. To build an API controller class, create a class named `ApiController`, as shown here:

```
@RestController
public class ApiController {
  private final EmployeeRepository repository;
  public ApiController(EmployeeRepository repository) {
    this.repository = repository;
  }
}
```

This API controller class be described as follows:

- `@RestController`: Spring's annotation to signal that this class doesn't process templates but instead that all outputs are directly serialized to the HTML response

- `EmployeeRepository`: We are injecting the repository we defined earlier in this section through constructor injection

The simplest thing of all is to return all the `Employee` records we have. This can easily be done by adding the following method to our `ApiController` class:

```
@GetMapping("/api/employees")
Flux<Employee> employees() {
  return repository.findAll();
}
```

This web method is quite simple:

- `@GetMapping`: Maps HTTP GET `/api/employees` calls to this method
- `Flux<Employee>`: Indicates that this returns one (or more) `Employee` records
- `repository.findAll()`: By using the prebuilt `findAll` method from Spring Data Commons' `ReactiveCrudRepository` interface, we already have a method that will fetch all the data

The previous chapter had a simple Java Map, which required some finagling to make it work reactively. Because `EmployeeRepository` extends `ReactiveCrudRepository`, it's already got the reactive types baked into its method's return types – no finagling needed!

This also means that coding an API-based POST operation can be done like this:

```
@PostMapping("/api/employees")
Mono<Employee> add(@RequestBody Mono<Employee>
  newEmployee) {
    return newEmployee.flatMap(e -> {
      Employee employeeToLoad =
        new Employee(e.getName(), e.getRole());
      return repository.save(employeeToLoad);
    });
}
```

This web method has the following features:

- `@PostMapping()`: Maps HTTP POST `/api/employees` calls to this method.
- `Mono<Employee>`: This method returns, at most, one entry.
- `@RequestBody(Mono<Employee>)`: This method will deserialize the incoming request body into an `Employee` object, but because it's wrapped as `Mono`, this processing only happens when the system is ready.
- `newEmployee.flatMap()`: This is how we access the incoming `Employee` object. Inside the `flatMap` operation, we actually craft a brand-new `Employee` object, deliberately dropping any `id` value provided in the input. This ensures a completely new entry will be made to the database.
- `repository.save()`: Our `EmployeeRepository` will execute a `save` operation and return `Mono<Employee>`, with a newly created `Employee` object inside. This new object will have everything, including a fresh `id` field.

If you're new to reactive programming, the ceremony of these preceding bullet points may be a little confusing. For example, why are we `flatMapping`? Mapping is usually used when converting from one type to another. In this situation, we're trying to map from an incoming `Employee` to a newly saved `Employee` type. So why don't we just map?

That's because what we got handed back from `save()` wasn't an `Employee` object. It was `Mono<Employee>`. If we had mapped over it, we'd have `Mono<Mono<Employee>>`.

> **flatMap is Reactor's golden hammer**
>
> A lot of times, when you're not sure what to do, or the Reactor APIs seem to be working against you, the secret is often `flatMap()`. All the Reactor types are heavily overloaded to support `flatMap` such that `Flux<Flux<?>>`, `Mono<Mono<?>>`, and every combination of them will nicely work out when you simply apply `flapMap()`. This also applies if you use Reactor's `then()` operator. Using `flatMap()` before you use `then()` will usually ensure the previous step is carried out!

The last step in putting together our reactive web app is to populate our Thymeleaf template, which we'll tackle in the next section.

Reactively dealing with data in a template

To finish things off, we need to create a `HomeController` class, like this:

```
@Controller
public class HomeController {
  private final EmployeeRepository repository;
  public HomeController(EmployeeRepository repository) {
    this.repository = repository;
  }
}
```

This class has some key aspects:

- `@Controller`: Indicates that this controller class is focused on rendering templates

- `EmployeeRepository`: Our beloved `EmployeeRepository` is also injected into this controller using **constructor injection**

With this in place, we can generate the web template served up at the root of our domain using this:

```
@GetMapping("/")
Mono<Rendering> index() {
  return repository.findAll() //
    .collectList() //
    .map(employees -> Rendering //
      .view("index") //
      .modelAttribute("employees", employees) //
      .modelAttribute
       ("newEmployee", new Employee("", ""))
      .build());
}
```

This is almost the same as the previous chapter's index() method except for the highlighted fragment:

- repository.findAll(): Instead of converting a map's values into Flux, EmployeeRepository already gives us one with its findAll() method

Everything else is the same.

Now, in order to process that form-backed Employee bean, we need a POST-based web method like this:

```
@PostMapping("/new-employee")
Mono<String> newEmployee(@ModelAttribute Mono<Employee>
  newEmployee) {
    return newEmployee //
      .flatMap(e -> {
        Employee employeeToSave =
          new Employee(e.getName(), e.getRole());
        return repository.save(employeeToSave);
      }) //
      .map(employee -> "redirect:/");
}
```

This is very similar to the previous chapter's `newEmployee()` method, except for the highlighted parts:

- `flatMap()`: As mentioned in the previous section, because `save()` returns a `Mono<Employee>`, we need to `flatMap` the results.

- The previous section also showed how we extract the name and role from the incoming `Employee` object but ignore any possible `id` value, since we're inserting a new entry. We then return the results of our repository's `save()` method.

- `map(employee -> "redirect:/")`: Here, we translate the saved `Employee` object into a redirect request.

Something that's important to point out, when compared to the previous chapter, is that in the preceding code, we split things up. In the previous chapter, we essentially mapped the incoming `Employee` object into a redirect request. That's because our mock database was non-reactive and only needed an imperative call to store the data.

Because our `EmployeeRepository` in this chapter is reactive, we need to split this up, with one operation focused on `save()`, followed by the next one, to convert that outcome to the redirect request.

Also, we had to use `flatMap` because the response from `save()` was wrapped inside a Reactor Mono class. Converting an employee to `"redirect:/"` doesn't involve any Reactor types, so simple mapping is all we need.

To get the `index.html` template, you can simply copy it from the previous chapter. Since it's the same as the previous chapter, there is no need to write it out here – no changes required! (Alternatively, grab it from the source code listing shown at the start of this chapter.)

And with that, we have a fully armed and operational reactive data store!

Summary

In this chapter, we learned about what it means to fetch data reactively. Based upon that, we selected a reactive data store and leveraged Spring Data to help manage our content. After hooking things into our reactive web controller, we then took a peek at R2DBC, a reactive driver for relational databases. And with that, we were able to build an introductory reactive Spring Boot application.

The same tactics used in earlier chapters for deployment will work just as well. In addition to that, many of the features we used throughout this book also work.

With all the things covered in this book, you should be prepared to take on your next (or perhaps current) project using Spring Boot 3.0. I sincerely hope Spring Boot 3.0 makes you as excited as I am about building new apps.

In the meantime, if you want to explore even more content regarding Spring Boot, check out the following resources:

- `http://bit.ly/3uSPLCz`: Check out my YouTube channel, loaded with free videos about Spring Boot

- `https://springbootlearning.com/medium`: I write weekly articles on Spring Boot as well as software engineering in general

- `https://springbootlearning.com/podcast`: I periodically post audio-only episodes where I either interview leaders from the Spring community or share morsels of knowledge about the world of Spring

- `https://twitter.com/springbootlearn`: If you want to connect, follow me on Twitter

Happy coding!

Index

Symbols

A

B

C

Packt.com

Subscribe to our online digital library for full access to over 7,000 books and videos, as well as industry leading tools to help you plan your personal development and advance your career. For more information, please visit our website.

Why subscribe?

- Spend less time learning and more time coding with practical eBooks and Videos from over 4,000 industry professionals

- Improve your learning with Skill Plans built especially for you

- Get a free eBook or video every month

- Fully searchable for easy access to vital information

- Copy and paste, print, and bookmark content

Did you know that Packt offers eBook versions of every book published, with PDF and ePub files available? You can upgrade to the eBook version at packt.com and as a print book customer, you are entitled to a discount on the eBook copy. Get in touch with us at customercare@packtpub.com for more details.

At www.packt.com, you can also read a collection of free technical articles, sign up for a range of free newsletters, and receive exclusive discounts and offers on Packt books and eBooks.

Other Books You May Enjoy

If you enjoyed this book, you may be interested in these other books by Packt:

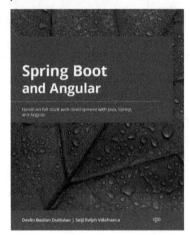

Spring Boot and Angular

Devlin Basilan Duldulao, Seiji Ralph Villafranca

ISBN: 978-1-80324-321-4

- Explore how to architect Angular for enterprise-level app development
- Create a Spring Boot project using Spring Initializr
- Build RESTful APIs for enterprise-level app development
- Understand how using Redis for caching can improve your application's performance
- Discover CORS and how to add CORS policy in the Spring Boot application for better security
- Write tests to maintain a healthy Java Spring Boot application
- Implement testing and modern deployments of frontend and backend applications

Full Stack Development with Spring Boot and React - Third Edition

Juha Hinkula

ISBN: 978-1-80181-678-6

- Make fast and RESTful web services powered by Spring Data REST
- Create and manage databases using ORM, JPA, Hibernate, and more
- Explore the use of unit tests and JWTs with Spring Security
- Employ React Hooks, props, states, and more to create your frontend
- Discover a wide array of advanced React and third-party components
- Build high-performance applications complete with CRUD functionality
- Harness MUI to customize your frontend
- Test, secure, and deploy your applications with high efficiency

Packt is searching for authors like you

If you're interested in becoming an author for Packt, please visit `authors.packtpub.com` and apply today. We have worked with thousands of developers and tech professionals, just like you, to help them share their insight with the global tech community. You can make a general application, apply for a specific hot topic that we are recruiting an author for, or submit your own idea.

Share Your Thoughts

Now you've finished *Learning Spring Boot 3.0*, we'd love to hear your thoughts! Scan the QR code below to go straight to the Amazon review page for this book and share your feedback or leave a review on the site that you purchased it from.

https://packt.link/r/1803233303

Your review is important to us and the tech community and will help us make sure we're delivering excellent quality content.

Download a free PDF copy of this book

Thanks for purchasing this book!

Do you like to read on the go but are unable to carry your print books everywhere?

Is your eBook purchase not compatible with the device of your choice?

Don't worry, now with every Packt book you get a DRM-free PDF version of that book at no cost.

Read anywhere, any place, on any device. Search, copy, and paste code from your favorite technical books directly into your application.

The perks don't stop there, you can get exclusive access to discounts, newsletters, and great free content in your inbox daily

Follow these simple steps to get the benefits:

1. Scan the QR code or visit the link below

https://packt.link/free-ebook/9781803233307

2. Submit your proof of purchase
3. That's it! We'll send your free PDF and other benefits to your email directly

www.ingramcontent.com/pod-product-compliance
Lightning Source LLC
Chambersburg PA
CBHW060532060326
40690CB00017B/3460

* 9 7 8 1 8 0 3 2 3 3 3 0 7 *